# Contemporary Coding Techniques and Applications for Mobile Communications

# OTHER TELECOMMUNICATIONS BOOKS FROM AUERBACH

# Contemporary Coding Techniques and Applications for Mobile Communications

*Onur Osman*
*Osman Nuri Uçan*

CRC Press
Taylor & Francis Group
Boca Raton London New York

CRC Press is an imprint of the
Taylor & Francis Group, an **informa** business
AN AUERBACH BOOK

CRC Press
Taylor & Francis Group
6000 Broken Sound Parkway NW, Suite 300
Boca Raton, FL 33487-2742

First issued in paperback 2019

ISBN-13: 978-1-4200-5461-3 (hbk)
ISBN-13: 978-0-367-38563-7 (pbk)

This book contains information obtained from authentic and highly regarded sources. Reasonable efforts have been made to publish reliable data and information, but the author and publisher cannot assume responsibility for the validity of all materials or the consequences of their use. The authors and publishers have attempted to trace the copyright holders of all material reproduced in this publication and apologize to copyright holders if permission to publish in this form has not been obtained. If any copyright material has not been acknowledged please write and let us know so we may rectify in any future reprint.

---
### Library of Congress Cataloging-in-Publication Data
---
Osman, Onur.
   Contemporary coding techniques and applications for mobile communications / Onur Osman and Osman Nuri Ucan.
     p. cm.
   ISBN 978-1-4200-5461-3 (alk. paper)
   1. Coding theory. 2. Mobile communication systems. I. Ucan, Osman Nuri. II. Title.

TK5102.92.O86 2007
621.3845'6--dc22                                          2007014521
---

**Visit the Taylor & Francis Web site at**
**http://www.taylorandfrancis.com**

**and the Auerbach Web site at**
**http://www.auerbach-publications.com**

To our children

Bahadir and Bengisu Uçan, and Emre Osman

# Contents

# Preface

*Contemporary Coding Techniques and Applications for Mobile Communications* covers both classical well-known coding techniques and contemporary codes and their applications.

The first four chapters include general information on mobile communications. In Chapter 5, AWGN, flat fading, time varying multipath, Wide Sense Stationary Uncorrelated Scattering Channels and MIMO channels are explained. The chapters following cover modern and applicable coding techniques.

In Chapter 6, channel equalization; semi-blind and blind equalization with adaptive filters such as RLS, LMS, EVA, Kalman, ML, and genetic algorithms, are summarized. Turbo codes, introduced in 1993, are given in Chapter 7, supported with clearly drawn figures; performances are derived for various channel models.

Achievement of both continuous phase property and error performance improvement is vital in all mobile communication systems. Thus, phase and frequency are combined, and the continuous phase frequency shift keying (CPFSK) is developed. In Chapters 8 and 9, new contemporary coding techniques with continuous phase property are investigated, including low-density parity check codes, time-diversity turbo trellis-coded modulation (TD-TTCM), multilevel turbo coded-continuous phase frequency shift keying (MLTC-CPFSK), turbo trellis-coded-continuous phase frequency shift keying (TTC-CPFSK), and low-density parity check coded-continuous phase frequency shift keying (LDPCC-CPFSK).

Today, digital image transmission is becoming common in cell phones. The critical point is how to optimize the combination of two main scientific subjects of electronics engineering—image processing and communications. Digital image processing is based on the neighborhood relationship of pixels. However, the transmitted signals that are the mapped form of these pixels should be independent from each other, based on the maximum likelihood theory. Thus, the most important problem to be solved in the following years is the contradiction between the pixel neighborhood necessity required for image processing and the independent interleaving needed for higher performance in digital communications. The last chapter explains some new approaches on the very popular subject of image transmission.

We thank the editors of Taylor & Francis Group who helped us improve our book.

We appreciate the continuous and valuable support we've received throughout our careers from our wives, Birsen U•an and Aygen Osman.

**Onur Osman and Osman Nuri U•an**
Istanbul, Turkey

# Authors

**Osman Nuri U•an** was born in Kars, Turkey, in January 1960. He received BSc, MSc, and PhD degrees from the Electronics and Communication Engineering Department, Istanbul Technical University (ITU) in 1985, 1988, and 1995, respectively. From 1986 to 1997, he worked as a research assistant in the same university. He worked as supervisor at TUBITAK-Marmara Research Center in 1998. He served as chief editor of the *Proceedings of the International Conference of Earth Science and Electronics* (ICESE), held in 2001, 2002, and 2003.

He is chief editor of Istanbul University's *Journal of Electrical and Electronics Engineering (IU-JEEE)*, cited by the Engineering Index (EI). He is professor and head of the Telecommunications Section, Electrical and Electronics Engineering Department, Istanbul University, where he is also head of the Biomedical Section of the Science Institute of IU and supervisor of the Telecommunication Institute of Turkey. His current research areas include information theory-turbo coding, image processing-cellular neural networks, wavelet, and Markov random fields applications on real geophysics data, and biomedical- and satellite-based two-dimensional data. He is author or coauthor of six books and two book chapters.

**Onur Osman** was born in Istanbul in September, 1973. He graduated from the Electrical Engineering Department of Istanbul Technical University in 1994. He received his MSc degree in electrical engineering in 1998 and subsequently his PhD degree from the Electrical and Electronics Engineering Department of Istanbul University. He worked with a telecommunication firm for a year as an R&D engineer and is now an assistant professor at Istanbul Commerce University. He is on the advisory board of the Istanbul University *Journal of Electrical and Electronics Engineering (IU-JEEE)*. His research area is information theory, channel coding, turbo coding, satellite communications, jitter performance of modulated systems, channel parameter estimation, neural networks, and image processing for biomedical and geophysics applications. He is the author or coauthor of five books and one book chapter.

# Chapter 1

# Introduction

## 1.1 Digital Communication Systems: An Overview

New improvements in wireless communication will increase the reliable exchange of information anywhere, anyplace, and anytime. Mobile communication environments are propagation characteristics, signal loss, multipath fading, and interference. Design countermeasures are design margins, diversity, coding, equalization, and error correction. Emerging wireless communications systems are based on contemporary coding, encryption, and channel estimation, where the main channel concept is frequency division, time division, spread spectrum.

The main scheme of a digital communication system, as in Figure 1.1, is composed of a source/channel encoder followed by a modulator. The modulated signals are passed through a noisy mobile channel. At the receiver, noisy signals are demodulated and decoded by source/channel decoders, and the estimated data is obtained.

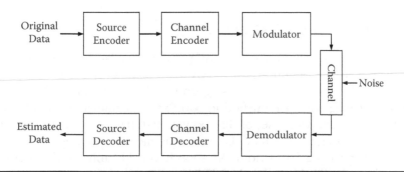

**Figure 1.1   General block diagram of a digital communication system.**

## 1.1.1  Source Coding

Original data is passed through a source encoder block. The main function of this block is to maximize entropy of the source. Entropy is symbolized by $H$, which shows average binary bits to represent each symbol. The replacement of the symbols with a binary representation is called source coding, which reduces the number of bits required to convey the information provided by original data. An instantaneous code is one that can be parsed as soon as the last bit of a codeword is received. An instantaneous code must satisfy the prefix condition. That is, no codeword may be a prefix of any other codeword. The best well-known source coding is Huffman coding. In image transmission, JPEG compression of digital images is based on this coding. It is assumed that both transmitter and receiver know the sequence set. The noiseless source coding theorem (also called Shannon's first theorem) states [1,2] that an instantaneous code can be found that encodes a source $x$ of entropy $H(x)$ with an average number of bits per symbol $B$ as

$$B \geq H(x) \tag{1.1}$$

## 1.1.2  Channel Coding

The task of source coding is to map the data source to an optimum number of symbols and to minimize error during their pass through the noisy channel. It is apparent that channel coding requires the use of redundancy. To detect or correct errors, the channel code sequence must be longer than the source sequence. To detect $n$ bit errors, a coding scheme requires the use of codewords with a Hamming distance of at least $n + 1$. In the same manner, to correct $n$ bit errors requires a coding scheme with at least a Hamming distance of $2n + 1$ between the codewords.

## 1.1.3  Modulation Systems

In classical analog communication; Amplitude Modulation (AM), Frequency Modulation (FM), Continuous Wave (CW) are commonly used. In AM, there is much to be gained in suppressing one of the sidebands and the carrier signal. Advantages are a reduction in band space, elimination of heterodyne whistles, and effective use of RF power. FM has a signal-to-noise-ratio advantage over amplitude modulation for the modulation index greater than 0.6 with a resulting bandwidth on speech considerably greater than that required for amplitude modulation. CW is a type of amplitude modulation [3].

In digital communication; frequency shift keying (FSK) and phase shift keying (PSK) are the main modulation types. FSK is a form of frequency modulation in which significant FM sidebands are generated. PSK is the constant amplitude modulation with the phase-shift keying property.

In 1982, trellis-coded modulation [4] combined convolutional coding and mapper block. It was the most powerful approach until 1993. Berrou et al. [5] introduced turbo codes in 1993. Turbo codes have become the favorite channel coding techniques. They have a complex decoding process but improve error performance up to 6 dB in AWGN.

### 1.1.4  Channel Capacity

Shannon's famous channel capacity theorem [1] is defined as

$$R \leq C \tag{1.2}$$

In a given channel with capacity $C$, there exists a code that will permit the error-free transmission across this channel at a rate $R$, as in Equation (1.2). Equality is achieved only when the signal-to-noise Ratio (SNR) tends to infinite.

## 1.2  Mathematical Fundamentals

In this subchapter, main mathematical fundamentals in communications are explained. The sampling theory and main concepts of the information theory will be briefly explained.

### 1.2.1  Sampling Theory

Signals in the real world, such as voices, are mainly analog signals. To process and keep these signals in computers, they must be converted to digital form. To convert a signal from continuous time to discrete time, a sampling process is used where the value of the signal is measured at certain intervals in time.

In the case of the signal having high-frequency components, a sample should be taken at a higher rate to avoid losing information in the signal. This is known as the Nyquist rate. The sampling theorem is as follows:

$$f_s \geq 2W \tag{1.3}$$

where $f_s$ sampling frequency and $W$ is the highest frequency to be sampled. In general, to preserve the full information in the signal, it is necessary to sample at twice the maximum frequency of the signal. If the analog signal is sampled lower than the Nyquist rate, it will exhibit a phenomenon called aliasing when converted back into a continuous time signal. In the case of aliasing, reconstruction of the original signal is impossible. Sometimes the highest frequency components of a signal may be noise. To prevent aliasing of these frequencies, preprocessing by low-pass filtering can be done before sampling the signal.

## 1.2.2 Information Theory

In information theory, an information source is a set of probabilities assigned to a set of outcomes. The information is carried not only by the outcome $\{x_i\}$, but by how uncertain it is [2,3]. A measure of the information contained in an outcome was introduced by Hartley in 1927. He defined the information $I(x_i)$ as

$$I(x_i) = \log_2\left(\frac{1}{P(x_i)}\right) = -\log_2\left(P(x_i)\right) \tag{1.4}$$

In Equation (1.4), it can be seen that an almost certain outcome contains less information, than one more unknown. As an example $I(x_i)$ becomes zero, meaning no information is conveyed, if its possibility is totally known or $P(x_i) = 1$.

The measure entropy, $H(\mathbf{X})$ defines the information content of the source $\mathbf{X}$ as a whole. It is the mean information provided by the source per source output or symbol as

$$H(\mathbf{X}) = \sum_i -P(x_i)\log_2\left(P(x_i)\right) = \sum_i P(x_i)I(x_i) \tag{1.5}$$

Entropy is maximized when all outcomes are equally likely, because entropy measures the average uncertainty of the source. If an information source $\mathbf{X}$ has $\mathbf{J}$ symbols, its maximum entropy is $\log_2(J)$ and this is obtained when all $\mathbf{J}$ outcomes are equally likely. Thus, for a $\mathbf{J}$ symbol source;

$$0 \leq H(X) \leq \log_2(J) \tag{1.6}$$

## References

[1] Shannon, C. E. A Mathematical Theory of Communication. *Bell System Technical Journal* 27:379–423 and 623–656, July and Oct. 1948.

[2] Slepian, D. (ed). *Key Papers in the Development of Information Theory.* New York: IEEE Press, 1974.

[3] Proakis, J., and D. Manolakis. *Digital Signal Processing: Principles, Algorithms, and Applications.* New York: Macmillan Publishing Company, 1992.

[4] Ungerboeck, G. Channel Coding with Multilevel/Phase Signals. *IEEE Trans. Inform. Theory* IT-28:55–67, 1982.

[5] Berrou, C., A. Glavieux, and P. Thitimajshima. Near Shannon Limit Error Correcting Coding and Decoding. *Proc. Intl. Conf. Comm.*:1064–1070, May 1993.

# Chapter 2

# Modulation Techniques

In digital communication, modulation block is one of the main elements [1–30]. The modulated signal is ready to transmit over high frequencies. Thus, it is easier to get rid of low-pass noise, such as the natural voice of living creatures. The electronic equipment gets smaller because of their high-frequency property. The most critical advantage of modulation is the antenna length, which is inversely proportional to frequency. As modulation frequency increases, the dimension of the antenna decreases.

## 2.1 Pulse Code Modulation

Pulse code modulation (PCM) is a digital representation of an analog signal defined within CCITT G.711 and ATT 43801. First, the magnitude of the analog signal is sampled at a certain frequency. Then, these samples are quantized to a series of symbols in a binary code. PCM is the standard of digital audio in computers and the compact disc format. It has been used in digital telephone systems and also is standard in digital video, for example, using ITU-R BT.601. Because PCM requires too much bandwidth, it is not generally used in video applications such as DVD (digital versatile disc) or DVR (digital video recording). Instead, compressed forms of digital audio are normally employed.

ITU-T G.711 is a standard to represent 8-bit compressed PCM samples for signals of voice using a sampling frequency of 8000 Hz. The G.711 encoder creates a 64 Kbps bit stream. This standard has two forms; A-Law and μ-Law. An A-Law G.711 PCM encoder converts 13-bit linear quantized

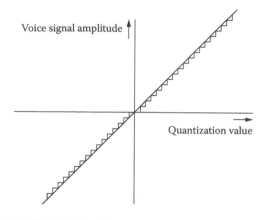

**Figure 2.1  Linear quantization.**

PCM samples (Figure 2.1) into 8-bit compressed PCM (logarithmic form) samples (Figure 2.2), and the decoder makes the conversion vice versa. μ-Law G.711 PCM encoder converts 14-bit linear PCM samples into 8-bit compressed PCM samples.

The μ-Law PCM is used in North America and Japan, and the A-Law is used in most other countries. The value used to represent the amplitude in both A-Law and μ-Law PCM is a number between 0 and +/– 127. Therefore, 8 bits are required to represent each sample (2 to the eighth power = 256).

The left-most bit (Bit 1) is transmitted first and is the most significant bit (MSB). This bit is known as the sign bit and is 1 for positive values and 0 for negative values in both PCM types. Bits 2 through 8 are inverted between A-Law and μ-Law PCM. Some codes of both PCM types are given

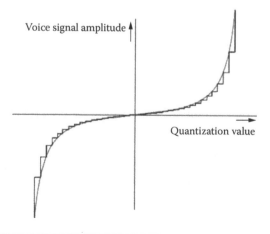

**Figure 2.2  Logarithmic quantization.**

**Table 2.1  PCM Codes of Numerical Values**

| Numerical Value | μ-Law 12345678 | A-Law 12345678 |
|:---:|:---:|:---:|
| +127 | 10000000 | 11111111 |
| +126 | 10000001 | 11111110 |
| +96 | 10011111 | 11100000 |
| +64 | 10111111 | 11000000 |
| +32 | 11011111 | 10100000 |
| 0 | 11111111 | 10000000 |
| 0 | 01111111 | 00000000 |
| −32 | 01011111 | 00100000 |
| −64 | 00111111 | 01000000 |
| −96 | 00011111 | 01100000 |
| −126 | 00000001 | 01111110 |
| −127 | 00000000 | 01111111 |

in Table 2.1. In A-Law, all even bits are inverted prior to transmission. The zero energy code of 00000000 is transmitted as 01010101.

The output of the μ-Law compander can be given as

$$output = \text{sgn}(x)\frac{\ln(1+u|x|)}{\ln(1+u)} \tag{2.1}$$

Here, $x$ is the normalized input between −1 and 1, $u$ is the compression parameter (generally used as 255), and $sgn(x)$ is the signum function of the normalized input. If $u$ tends to 0, companding becomes linear. The output of the A-Law compander can be given as

$$output = \begin{cases} \text{sgn}(x)\dfrac{A|x|}{1+\ln(A)} & 0 \le |x| < 1/A \\[3mm] \text{sgn}(x)\dfrac{1+\ln(A|x|)}{1+\ln(A)} & 1/A \le |x| \le 1 \end{cases} \tag{2.2}$$

Here, $A$ is the compression parameter, generally used as 87.6. If $A$ converges to 1, companding gets linear.

As an example, PCM encoding with 4-bit linear quantization is demonstrated in Figure 2.3 and Table 2.1. First, the input signal is sampled. When the voice waveform is sampled, a series of short pulses are produced, each

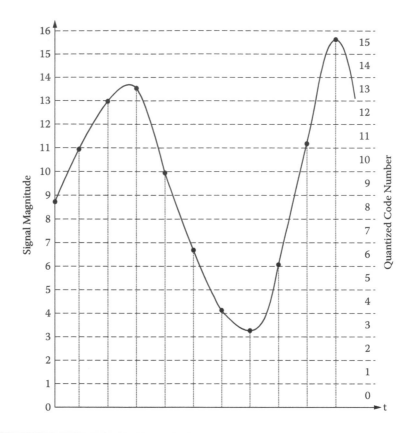

**Figure 2.3   Pulse code modulation (PCM): An example.**

representing the amplitude of the waveform at the specific instant of sampling. This process is called pulse amplitude modulation (PAM). Then, these values are quantized using a linear or nonlinear compander. In Table 2.2, linear quantization outputs are given.

To calculate the quantization noise, the quantization process can be modeled as

$$\tilde{x}_k = x_k + \epsilon_k \tag{2.3}$$

where $\tilde{x}_k$ is the quantized value of $x_k$ and $\epsilon_k$ is the quantization error. The probability density function (pdf) of the quantization noise is assumed to be uniform, while a linear quantizer is used, and given as

$$p(\epsilon) = \frac{1}{\Delta} \qquad -\frac{\Delta}{2} \le \epsilon \le \frac{\Delta}{2} \tag{2.4}$$

### Table 2.2 PCM with Linear Quantization

| PAM value | Quantized code number | PCM code |
|---|---|---|
| 8.6 | 8 | 1000 |
| 10.9 | 10 | 1010 |
| 12.8 | 12 | 1100 |
| 13.3 | 13 | 1101 |
| 9.9 | 9 | 1001 |
| 6.6 | 6 | 0110 |
| 4.1 | 4 | 0100 |
| 3.3 | 3 | 0011 |
| 7.1 | 7 | 0111 |
| 12.2 | 12 | 1100 |
| 15.5 | 15 | 1111 |

*Note:* PCM = pulse code modulation, PAM = pulse amplitude modulation.

where each step of the quantizer is

$$\Delta = \frac{1}{2^q}.$$

Here, $q$ represents the number of bits of the quantizer. The mean square quantization error is [30]

$$E(\epsilon^2) = \int_{-\Delta/2}^{\Delta/2} \epsilon^2 p(\epsilon^2) d\epsilon$$

$$= \frac{\Delta^2}{12} = \frac{1}{12 \times 2^{2q}}$$

(2.5)

The quantization noise in decibels is

$$10\log\left(\frac{1}{12 \times 2^{2q}}\right) = -6q - 10.8 \; dB.$$

## 2.2 Delta Modulation

In delta modulation, each signal sample is coded with a single bit, which is determined from the difference of the previous sample's value and the current one (Figure 2.4). The bit specifies whether or not the new sample is higher than the previous one. Therefore, the information only notifies the variation, but its contents do not indicate the value of sampled data. The resulting bit stream can be drawn as a staircase that approximates to the original analog signal; 1 and 0 being represented as the ascent and descent of the stair, respectively.

For example, if a standard 4-KHz voice channel is considered, at least 8000 samples per second should be needed to preserve the original signal. The signal can be codified with 8 Kbps using delta modulation. If we use PCM with 256 levels (with an 8-bit quantizer), the same signal could occupy 8000 × 8 bps or 64 Kbps.

It is clear that delta modulation needs to transmit a lot less data than PCM. But, delta modulation doesn't provide sufficient quality. Although it might work for monotonous voices, it doesn't adapt well to signals with sudden changes. Conversely, PCM with 256 levels provides very good quality.

Because the delta modulator approximates a waveform $\tilde{s}_n$ by a linear staircase function, the waveform $\tilde{s}_n$ changes slowly relative to the sampling rate. This requirement implies that waveform $\tilde{s}_n$ must be oversampled.

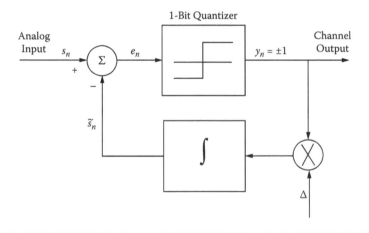

**Figure 2.4   Delta modulator.**

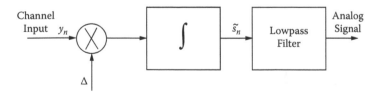

**Figure 2.5   Delta demodulator.**

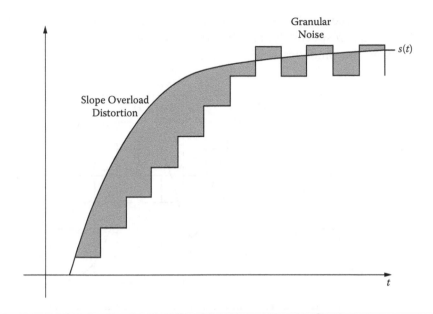

**Figure 2.6   Slope overload distortion and granular noise.**

Oversampling means that the signal is sampled faster than the Nyquist rate. That is, in the case of delta modulation, the sampling rate will be much higher than the minimum rate of twice the bandwidth. To obtain an accurate prediction of the next input, delta modulation requires oversampling. Because each encoded sample contains a relatively small amount of information, delta modulation systems require higher sampling rates than PCM systems. There are two types of distortion, as shown below, which limit the performance of the delta modulation encoder at any given sampling rate (Figures 2.5 and 2.6) [2].

- ▪ Slope overload distortion: This type of distortion occurs, if the step size of delta is too small to follow portions of the waveform that have a steep slope. It can be reduced by increasing the step size.
- ▪ Granular noise: This type of noise takes place if the step size is too large in parts of the waveform having a small slope. Granular noise can be reduced by decreasing the step size.

Even for an optimized step size, the performance of the delta modulator encoder may still be less satisfactory. This problem can be solved by employing a variable step size that adapts itself to the short-term characteristics of the source signal. The step size should be increased when the waveform has a steep slope and decreased when the waveform has a relatively gradual slope. This strategy is called adaptive delta modulation (ADM).

As an example, in Figures 2.7 and 2.8, delta modulation is applied for Δ = 1V and 2V, respectively, and binary representations of the signal are given below the timelines.

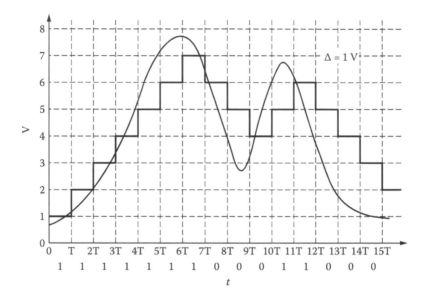

**Figure 2.7    Delta modulation for Δ = 1V.**

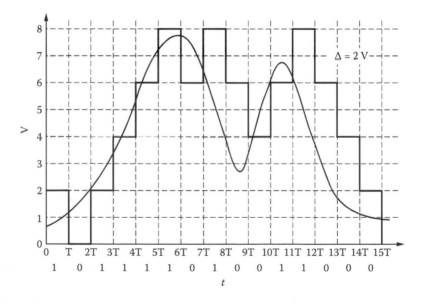

**Figure 2.8    Delta modulation for Δ = 2V.**

## 2.3  Adaptive Delta Modulation

In delta modulation, slope overloaded distortion and granular noise can be decreased by keeping the step size small and sampling at a rate many times higher than the Nyquist rate. This solution entails a sacrifice in bandwidth. Whereas, delta modulation is chosen because of its smaller need of bits to reform the source signal compared to PCM. A more effective way to reduce the slope overloaded distortion and the granular noise is to make the step size of delta modulation variable. This modulation method is called adaptive delta modulation (ADM).

A block diagram of ADM is shown in Figure 2.9. The step size of adaptive delta modulation is adjusted based on the formula below.

$$\Delta_n = \begin{cases} \Delta_{n-1} \times C & \textit{if } y_n = y_{n-1} \\ \\ \Delta_{n-1} / C & \textit{if } y_n = -y_{n-1} \end{cases} \qquad (2.6)$$

The step size is varied to adopt both steep and gradual slopes. The constant $C > 1$ determines the variation rate of the step size. If $C$ is close to 1, less change in step size would be observed. Otherwise, the value of the step size varies very much. In Figure 2.10, ADM is shown for $C = 2$.

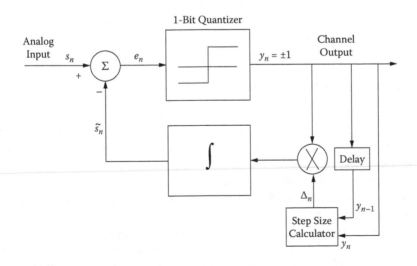

**Figure 2.9  Adaptive delta modulator.**

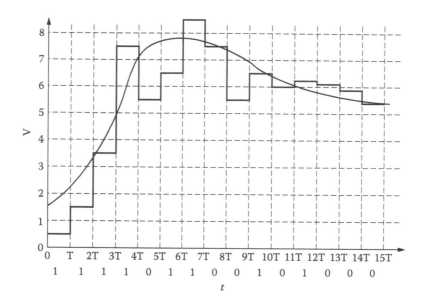

**Figure 2.10   Adaptive delta modulation for $C = 2$.**

## 2.4   Line Coding

In baseband transmissions, to send the binary logic data over the wire it must be coded into an electrical signal. Line coding is the method of selecting pulses to represent binary data. In general, there are four different main signaling methods: unipolar, polar, dipolar, and bipolar alternate mark inversion signaling.

### 2.4.1   Unipolar Signaling

In unipolar signaling, one binary symbol, digital zero, is represented by the absence of a pulse (i.e., a space) and the other binary symbol, digital one, is represented by a pulse (i.e., mark). There are two types of unipolar signalings: non-return to zero (NRZ) and return to zero (RZ). In NRZ, there is no variation in the signal during the time slot. However, RZ pulses fill only the first half of the time slot (Figure 2.11). The second half of the signal returns to zero.

Both NRZ and RZ unipolar signals have disadvantages. Because these signals have a DC level represented in their spectra by a line at 0 Hz (Figure 2.11 [30]). If these signals are transmitted over links with transformer- or capacitor-coupled (AC) repeaters, the line at 0 Hz is removed

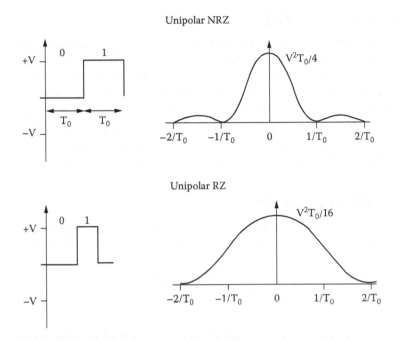

**Figure 2.11   Unipolar RZ and NRZ signals and their spectra.**

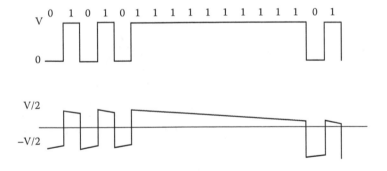

**Figure 2.12   Input and distorted output unipolar NRZ signal because of AC coupling.**

and the signal becomes a polar format. Because both RZ and NRZ signal spectra are non-zero at 0 Hz then the transmitted pulse shapes are distorted because of AC coupling. Therefore, when long strings of zeros or ones are present, a DC or baseline wander occurs (Figure 2.12 [30]).

## 2.4.2 Polar Signaling

Polar signaling refers to a line code in which binary ones and zeros are represented by pulses and opposite pulses respectively (Figure 2.13 [30]). The frequency spectra of polar RZ and NRZ signals have identical shapes with unipolar RZ and NRZ signals. Polar and unipolar signals have the same bandwidth requirements and if they are transmitted over AC-coupled lines, they suffer the same distortion effects. Conversely, polar signaling has a significant power advantage over unipolar signaling because there is no DC level. Therefore, its bit error ratio is better for the same signaling power.

## 2.4.3 Dipolar Signaling

In dipolar signaling, as shown in Figure 2.14 [30], the symbol interval is split into positive and negative pulses each with a width of $T_0/2$. Dipolar signaling has a spectral null at 0 Hz. Therefore, it is very suitable to AC-coupled transmission lines. Manchester coding is used for magnetic recording and ethernet local area networks.

## 2.4.4 Bipolar Alternate Mark Inversion (AMI) Signaling

Three different voltage levels (+V, 0, −V) are used in bipolar signaling to represent binary symbols. Therefore, it is called pseudo-ternary line coding.

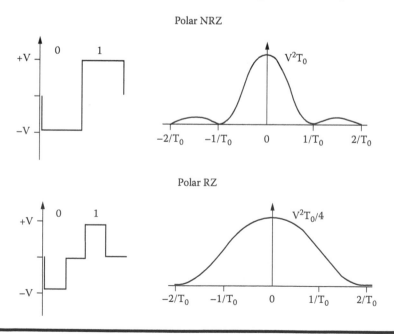

**Figure 2.13  Polar RZ and NRZ signals and their spectra.**

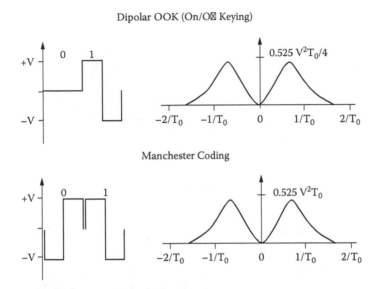

**Figure 2.14  Dipolar on/off keying and Manchester coding, and their spectra.**

Zeros are always represented by zero voltage level (i.e., absence of a pulse) and ones are represented alternately by voltage levels of +V and –V (Figure 2.15 [30]). Because the mark voltage level is alternating, the bipolar spectrum has a null at 0 Hz. Therefore, it is suitable to AC-coupled transmission lines. Moreover, the alternating mark voltage gives bipolar signaling a single error detection capability.

## 2.4.5  Coded Mark Inversion

In coded mark inversion signaling, positive and negative amplitude levels represent a digital 0 and the constant amplitude level represents a digital 1. It is a type of polar NRZ code as shown in Figure 2.16 [30]. In the case of several successive ones the amplitude level is inverted with each pulse. Therefore, CMI is a combination of dipolar signaling and NRZ AMI.

## 2.4.6  nBmT Coding

In *n*B*m*T line coding, *n* binary symbols are mapped into *m* ternary symbols. As an example, a 4B3T coding table is given in Table 2.3. 4B3T is used for the ISDN BRI interface. In this method, three pulses represent four binary bits. It uses three states: +, 0, and – to represent positive pulse, no pulse, and negative pulse, respectively. Here, with 4 bits we have 16 (=$2^4$) possible input combinations to represent using $3^3 = 27$ output combinations. 000 is not used to avoid long periods without a transition.

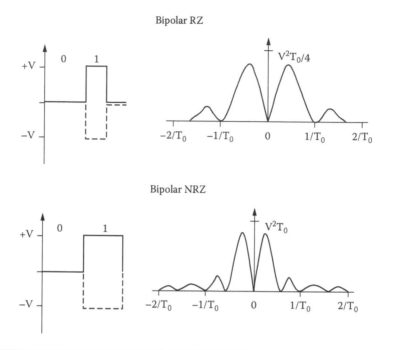

**Figure 2.15 Bipolar RZ and NRZ signals and their spectra.**

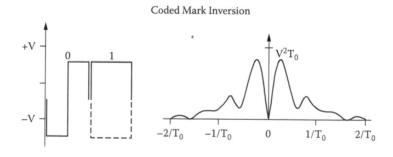

**Figure 2.16 Coded mark inversion and its spectra.**

Each 4-bit block of binary data is coded into three ternary symbols according to Table 2.3. There should be more than one ternary symbol for each bit block. Only one of them is given in Table 2.3. The decoding process is executed in the reverse manner of coding. Each block of 4 bits is obtained from the conversion of the received blocks of three ternary symbols. The decoding is done according to the formulas given in Table 2.4. The left-hand symbol of each block (both binary and ternary)

**Table 2.3   4B3T Coding Table**

| Binary Block | | | | S1 | | | S2 | | | S3 | | | S4 | | |
|---|---|---|---|---|---|---|---|---|---|---|---|---|---|---|---|
| 0 | 0 | 0 | 1 | 0 | − | + | 0 | − | + | 0 | − | + | 0 | − | + |
| 0 | 1 | 1 | 1 | − | 0 | + | − | 0 | + | − | 0 | + | − | 0 | + |
| 0 | 1 | 0 | 0 | − | + | 0 | − | + | 0 | − | + | 0 | − | + | 0 |
| 0 | 0 | 1 | 0 | + | − | 0 | + | − | 0 | + | − | 0 | + | − | 0 |
| 1 | 0 | 1 | 1 | + | 0 | − | + | 0 | − | + | 0 | − | + | 0 | − |
| 1 | 1 | 1 | 0 | 0 | + | − | 0 | + | − | 0 | + | − | 0 | + | − |
| 1 | 0 | 0 | 1 | + | − | + | + | − | + | + | − | + | − | − | − |
| 0 | 0 | 1 | 1 | 0 | 0 | + | 0 | 0 | + | 0 | 0 | + | − | − | 0 |
| 1 | 1 | 0 | 1 | 0 | + | 0 | 0 | + | 0 | 0 | + | 0 | − | 0 | − |
| 1 | 0 | 0 | 0 | + | 0 | 0 | + | 0 | 0 | + | 0 | 0 | 0 | − | − |
| 0 | 1 | 1 | 0 | − | + | + | − | + | + | − | − | + | − | − | + |
| 1 | 0 | 1 | 0 | + | + | − | + | + | − | + | − | − | + | − | − |
| 1 | 1 | 1 | 1 | + | + | 0 | 0 | 0 | − | 0 | 0 | − | 0 | 0 | − |
| 0 | 0 | 0 | 0 | + | 0 | + | 0 | − | 0 | 0 | − | 0 | 0 | − | 0 |
| 0 | 1 | 0 | 1 | 0 | + | + | − | 0 | 0 | − | 0 | 0 | − | 0 | 0 |
| 1 | 1 | 0 | 0 | + | + | + | − | + | − | − | + | − | − | + | − |

**Table 2.4   4B3T Decoding Table**

| Ternary Block | | | | | | | | | Binary Block | | | |
|---|---|---|---|---|---|---|---|---|---|---|---|---|
| 0 | 0 | 0 | + | 0 | + | 0 | − | 0 | 0 | 0 | 0 | 0 |
| 0 | + | − | | | | | | | 0 | 0 | 0 | 1 |
| + | − | 0 | | | | | | | 0 | 0 | 1 | 0 |
| 0 | 0 | + | − | − | 0 | | | | 0 | 0 | 1 | 1 |
| − | + | 0 | | | | | | | 0 | 1 | 0 | 0 |
| 0 | + | + | − | 0 | 0 | | | | 0 | 1 | 0 | 1 |
| − | + | + | − | − | + | | | | 0 | 1 | 1 | 0 |
| − | 0 | + | | | | | | | 0 | 1 | 1 | 1 |
| + | 0 | 0 | 0 | − | − | | | | 1 | 0 | 0 | 0 |
| + | − | + | − | − | − | | | | 1 | 0 | 0 | 1 |
| + | + | − | + | − | − | | | | 1 | 0 | 1 | 0 |
| + | 0 | − | | | | | | | 1 | 0 | 1 | 1 |
| + | + | + | − | + | − | | | | 1 | 1 | 0 | 0 |
| 0 | + | 0 | − | 0 | − | | | | 1 | 1 | 0 | 1 |
| 0 | + | − | | | | | | | 1 | 1 | 1 | 0 |
| + | + | 0 | 0 | 0 | − | | | | 1 | 1 | 1 | 1 |

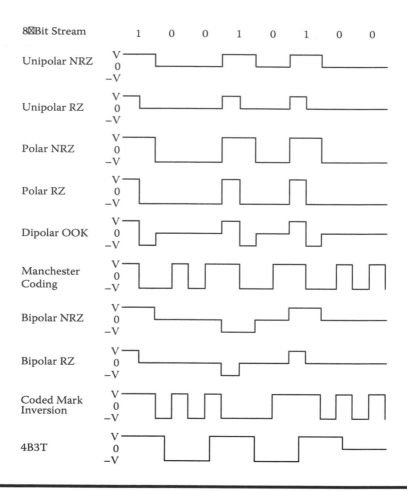

**Figure 2.17 Various line coding methods for an 8-bit sequence.**

is the first bit and the right hand is the last as shown in the encoding table. During deactivation, the ternary block "0 0 0" is received and it is decoded to binary "0 0 0 0".

The various line coding methods for an 8-bit sequence are given in Figure 2.17.

## 2.5 Amplitude Modulation

Amplitude modulation (AM) is a technique of impressing information data onto a radio carrier wave, which is most commonly used in communication. AM works by altering the amplitude of the transmitted signal according to the information being sent. In Figure 2.18, an AM signal is given. Here changes in the signal amplitude reflect the sounds

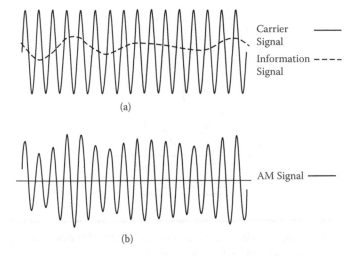

(a)

(b)

**Figure 2.18** **(a) The carrier and information signal and (b) amplitude modulation (AM) signal.**

to be reproduced by a speaker, or to specify the light intensity of television pixels.

The mathematical formula of amplitude modulation can be given as

$$x_{AM}(t) = A_c\left[1 + mx(t)\right]\cos(2\pi f_c t) \qquad (2.7)$$

where $m \in [0, 1]$ is the modulation index, $x(t) \in [-1, 1]$ is the original message, $f_c$ is the carrier frequency, $A_c$ is the carrier amplitude, and $x_{AM}(t)$ is the amplitude modulated signal. The Fourier transformation of $x_{AM}(t)$ is given below and the spectrum of the signal is shown in Figure 2.19.

$$x_{AM}(f) = \frac{A_c}{2}\left[\delta(f - f_c) + \delta(f + f_c) + mx(f - f_c) + mx(f + f_c)\right] \quad (2.8)$$

In AM, up to 33 percent of the overall signal power is contained in the sidebands when a carrier is amplitude-modulated with a perfect sine wave. Unfortunately, the remaining part of the signal power is contained in the carrier, and that does not contribute to the transfer of data. If more complex modulating signals are considered, such as video, music, or voice, the sidebands generally contain 20 to 25 percent of the overall signal power. Thus the carrier consumes 75 to 80 percent of the power that is not related to the transmission of data. This makes AM an inefficient modulation technique. If the modulating data input amplitude is increased

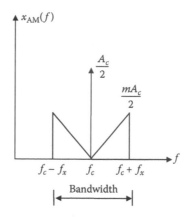

**Figure 2.19 Right-hand side of a frequency spectrum of modulated signal, $X_c(f)$ with $f_x$ maximum modulating signal frequency.**

beyond these limits, the signal will become distorted, and will occupy a much greater bandwidth than it should. This can be the result of signal interference on nearby frequencies, which is called *overmodulation*.

The single method of increasing the transmitter efficiency is to remove (suppress) the carrier from the AM signal. This results in a reduced-carrier transmission or double-sideband suppressed carrier (DSBSC) signal. A suppressed-carrier mode is three times more power-efficient than the traditional DSB-AM. If the carrier is only partially suppressed, a double-sideband reduced carrier (DSBRC) signal occurs. DSBSC and DSBRC signals need their carrier to be regenerated and to be demodulated using conventional techniques. Greater efficiency can be achieved by completely suppressing both the carrier and one of the sidebands, but this results in increased transmitter and receiver complexity. This is called single-sideband modulation, which is widely used in amateur radio because of its power and bandwidth efficiency.

## 2.6 Frequency Modulation

In contrast to amplitude modulation in which only the frequency of the carrier is varied while its amplitude remains constant, frequency modulation (FM) is a modulation technique that represents the information with variations in the instantaneous frequency of a carrier wave (Figure 2.20). The carrier frequency is varied according to the changes in the amplitude of an input signal in analog applications. In digital applications, frequency shift keying (FSK) is used instead of FM, shifting the carrier frequency among a set of discrete values.

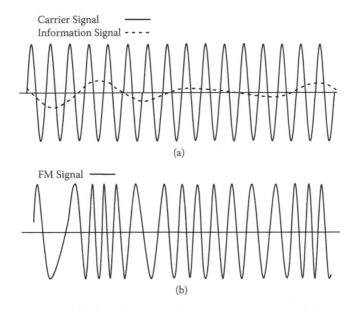

**Figure 2.20** **(a) Carrier and information signal and (b) FM signal.**

FM is commonly used at VHF radio frequencies to broadcast music, speech, and analog TV sound. In commercial and amateur radio systems, a narrowband form is used for voice communications. The type of FM used in broadcast, such as radio or television, is generally called wide-FM (W-FM). In two-way radio communication, bandwidth conservation becomes important and thus, narrowband narrow-FM (N-FM) is used.

In W-FM, wider bandwidth is required than amplitude modulation using an equivalent modulating signal, but FM also makes the signal more robust against noise and interference. FM is also more robust against amplitude fading phenomena because the information carried is in the frequency of the signal not in its amplitude. Therefore, FM was chosen as the modulation standard for high-frequency and high-quality radio transmission.

Assume that $x(t) \in [-1, 1]$ is the information data (generally music or speech) and $x_c(t)$ is the carrier signal.

$$x_c(t) = A \cos(2\pi f_c t) \qquad (2.9)$$

where, $f_c$ is the carrier frequency and $A$ is an arbitrary amplitude. The carrier is modulated by the signal given below.

$$x_{FM}(t) = A \cos \left( 2\pi f_c + \Delta f \int_0^t x(t)dt \right) \qquad (2.10)$$

where $\Delta f$ is the frequency deviation, which represents the maximum shift away from $f_c$ in one direction. The modulation index of FM is defined as

$$\beta = \frac{\Delta f}{f_m} \qquad (2.11)$$

where $f_m$ is the maximum modulating frequency of $x(t)$. The bandwidth of an FM signal may be given by using Carson's rule,

$$W = 2 f_m (\beta + 1) \qquad (2.12)$$

The number of side band pairs is 1 for $\beta < 1$ and $\beta + 1$ for other values. The audio spectrum range is between 20 and 20,000 Hz, but FM radio limits the upper modulating frequency to 15 KHz. If FM transmitters use a maximum modulation index of 5.0, the resulting bandwidth is 180 KHz.

## 2.7 Phase-Shift Keying

Phase-shift keying (PSK) is a constant amplitude modulation type with symbols constellation of phase differences-delays. The simple PSK is binary PSK.

### 2.7.1 Binary Phase-Shift Keying

Binary phase-shift keying (BPSK) is a digital modulation scheme that conveys binary data by modulating the phase of a carrier signal. The modulating signal represents one bit in the transmission period. The information (1 or 0) is in the phase of the signal. In BPSK, two different signals can be sent because the information bit is to be one or zero. Therefore, two different phases are used for transmitting waveforms, and also these phases are chosen to obtain the maximum distance between them. The signal constellation of BPSK is shown in Figure 2.21. It uses two phases that are separated by 180° and so can also be termed 2-PSK. It is not important where the constellation points are positioned, and in this figure they are shown on the real axis (Inphase), at 0° and 180°, and the values on the imaginary axis (quadrature-phase) are zero.

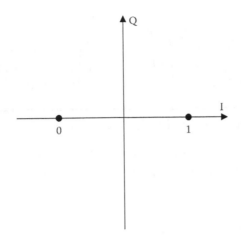

**Figure 2.21  Binary phase-shift keying (BPSK) signal constellation.**

This modulation is the most robust of all the PSKs because it requires serious distortion to make the demodulator reach an incorrect decision. However, it is only able to modulate at 1 bit/symbol and so is unsuitable for high data-rate applications. For baseband communication, the binary values are passed through a transformation of

$$X(t) = 2x(t) - 1 \tag{2.13}$$

For bandpass communication, baseband values of binary data (1 or −1) are multiplied by the carrier signal. Thus, the forms of the transmitting signal are shown in Table 2.5.

**Table 2.5  BPSK Binary Data and Mapped Signals**

| Binary Information Data | Baseband Data | Bandpass Signal |
|:---:|:---:|:---:|
| 1 | 1 | $s_1(t) = \sqrt{\dfrac{2E_b}{T}} \cos(2\pi f_c t)$ |
| 0 | 1 | $s_0(t) = -\sqrt{\dfrac{2E_b}{T}} \cos(2\pi f_c t)$ |

*Note:* BPSK = Binary phase-shift keying.

The signal waveform of a bit stream is shown in Figure 2.22 for BPSK.

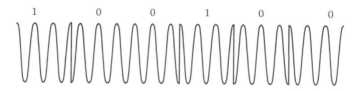

**Figure 2.22  Bandpass signal of a bit stream modulated by BPSK.**

Because there is only one bit per symbol, the symbol error rate (SER) is the same as bit error rate (BER). The bit error rate (BER) of BPSK in AWGN can be calculated as

$$P_b = \frac{1}{2}\left[1 - erf\left(\frac{E_b}{N_0}\right)^{1/2}\right] \qquad (2.14)$$

## 2.7.2  M-ary PSK

M-ary PSK is used to transmit more than one bit within the same symbol. These symbols consist of 2, 3 or 4 bits but 8PSK is usually the highest order PSK constellation deployed. With more than eight phases, the error rate becomes too high and there are better, though more complex, modulations available such as quadrature amplitude modulation (QAM). Although any number of phases may be used, the fact that the constellation must usually deal with binary data means that the number of symbols is usually a power of two. Assume that the encoder output has 2 bits, so four different signals are needed to represent these bits. These four symbols can be located with 90° differences on the complex plane of the mapper as shown in Figure 2.23.

All 2-bit sequences and their complex values and signals are given in Table 2.6.

If the output of the encoder has 3 bits, eight different signals with 45° phase differences are needed as shown in Figure 2.24.

All 3-bit sequences and their complex values and signals are given in Table 2.7.

A simple and good approximation of symbol error probability ($P_e$) for M-ary PSK (for $M \geq 4$) can be given as

$$P_e = 1 - erf\left[\sin(\pi / M)\left(\frac{E_s}{N_0}\right)^{1/2}\right] \qquad (2.15)$$

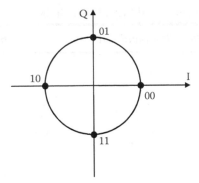

**Figure 2.23   A 4-ary PSK signal constellation.**

**Table 2.6   Complex Values and Signals for 4PSK**

| Binary Information Data | Baseband Data | Bandpass Signal |
| --- | --- | --- |
| 00 | 1 | $\sqrt{\dfrac{2E_s}{T}}\cos(wt)$ |
| 01 | $J$ | $\sqrt{\dfrac{2E_s}{T}}\sin(wt)$ |
| 10 | 1 | $-\sqrt{\dfrac{2E_s}{T}}\cos(wt)$ |
| 11 | $-j$ | $-\sqrt{\dfrac{2E_s}{T}}\sin(wt)$ |

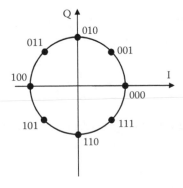

**Figure 2.24   An 8-ary PSK signal constellation.**

**Table 2.7  Complex Values and Signals for 8PSK**

| Binary Information Data | Baseband Data | Bandpass Signal |
|:---:|:---:|:---:|
| 000 | $1$ | $\sqrt{\dfrac{2E_s}{T}}\cos(wt)$ |
| 001 | $\dfrac{\sqrt{2}}{2}+j\dfrac{\sqrt{2}}{2}$ | $\dfrac{\sqrt{2}}{2}\sqrt{\dfrac{2E_s}{T}}\cos(wt)+\dfrac{\sqrt{2}}{2}\sqrt{\dfrac{2E_s}{T}}\sin(wt)$ |
| 010 | $j$ | $\sqrt{\dfrac{2E_s}{T}}\sin(wt)$ |
| 011 | $-\dfrac{\sqrt{2}}{2}+j\dfrac{\sqrt{2}}{2}$ | $-\dfrac{\sqrt{2}}{2}\sqrt{\dfrac{2E_s}{T}}\cos(wt)+\dfrac{\sqrt{2}}{2}\sqrt{\dfrac{2E_s}{T}}\sin(wt)$ |
| 100 | $-1$ | $-\sqrt{\dfrac{2E_s}{T}}\cos(wt)$ |
| 101 | $-\dfrac{\sqrt{2}}{2}-j\dfrac{\sqrt{2}}{2}$ | $-\dfrac{\sqrt{2}}{2}\sqrt{\dfrac{2E_s}{T}}\cos(wt)-\dfrac{\sqrt{2}}{2}\sqrt{\dfrac{2E_s}{T}}\sin(wt)$ |
| 110 | $-j$ | $-\sqrt{\dfrac{2E_s}{T}}\sin(wt)$ |
| 111 | $\dfrac{\sqrt{2}}{2}-j\dfrac{\sqrt{2}}{2}$ | $\dfrac{\sqrt{2}}{2}\sqrt{\dfrac{2E_s}{T}}\cos(wt)-\dfrac{\sqrt{2}}{2}\sqrt{\dfrac{2E_s}{T}}\sin(wt)$ |

The bit error probability is

$$P_b = \frac{P_e}{\log_2 M} \tag{2.16}$$

Since the energy of all symbols in M-ary PSK is identical, so the average energy per bit can be given as

$$E_b = \frac{E_s}{\log_2 M} \tag{2.17}$$

Thus, the approximate bit error probability with respect to bit energy and $M$ is [30]

$$P_b = \frac{1}{\log_2 M}\left\{1 - erf\left[\sin(\pi/M)\sqrt{\log_2 M}\left(\frac{E_b}{N_0}\right)^{1/2}\right]\right\} \tag{2.18}$$

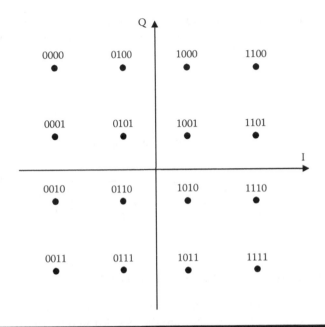

**Figure 2.25    A 16-QAM signal constellation.**

## 2.7.3  *Quadrature Amplitude Modulation*

Quadrature amplitude modulation (QAM) is a modulation scheme that conveys data by changing the amplitude of two carrier signals. If the encoder output has more than 3 bits, the Euclidean distance between the symbols in MPSK system becomes less. Therefore, QAM is used when lots of bits are needed to be transmitted in one symbol. In Figure 2.25, a 16-QAM signal constellation is shown.

A simple approximation for the probability of symbol error for M-ary QAM signaling in AWGN is [31]

$$P_s = 2\left\{\frac{M^{1/2}-1}{M^{1/2}}\right\}\left[1 - erf\sqrt{\frac{3}{2(M-1)}\left(\frac{[E]}{N_0}\right)^{1/2}}\right] \qquad (2.19)$$

where $[E]$ is the average energy per QAM symbol. For equiprobable rectangle pulse symbols,

$$[E] = \frac{1}{3}\left(\frac{\Delta V}{2}\right)^2 (M-1)T_0 \qquad (2.20)$$

where $\Delta V$ is the amplitude separation between adjacent inphase or quadrature phase levels and $T_0$ is the period of the symbol. The average energy per bit can be computed as

$$E_b = \frac{[E]}{\log_2 M}$$

(2.21)

Consequently, the approximate bit-error probability is [30]

$$P_b = \frac{2}{\log_2 M} \left\{ \frac{M^{1/2} - 1}{M^{1/2}} \right\} \left[ 1 - erf \sqrt{\frac{3 \log_2 M}{2(M-1)} \left( \frac{E_b}{N_0} \right)^{1/2}} \right]$$

(2.22)

## 2.8 Continuous Phase Modulation

To transmit the symbols in digital phase modulation schemes, the phase of the carrier is used. PSK and FSK are the best known modulation methods for digital communication systems. In PSK, the phase of the signal is chosen from among a set of a finite number of values according to the input symbol. Contrary to PSK, FSK changes the frequency of the signal nearby the carrier frequency to transmit different symbols [6]. Neither FSK nor PSK can provide the phase continuity. Therefore, high frequency components occur and thus, wide-band communication is needed.

In recent decades, the most important advance in coding theory is the development of coding techniques that enhance the efficiency of bandwidth. Modulation schemes also provide extra coding gain preserving both signal energy and bandwidth. The idea of preserving the energy and bandwidth by using coding was introduced by Shannon [7], but the studies on this subject accelerated after 1975.

In the beginning, it was thought that bandwidth should be increased for energy efficiency. The bandwidth criterion was that the 99 percent of the energy must be transmitted through the bandwidth.

When PSK and QAM are combined with error correcting codes (ECC) they need less energy but higher bandwidth. Continuous phase modulation (CPM) with constant envelope is the most important subset when both energy and bandwidth efficient modulation techniques are considered.

This modulation technique not only has narrow main spectra and very low side spectra, but also has better error performance for the same $E_b/N_0$ according to PSK or QAM. The method, in which phase modulation function is changed relative to the frequency with constant amplitude, is called continuous-phase frequency shift keying (CPFSK) [12–15].

The CPM modulator can be conceived as a finite state machine defining the states at the beginning of the signaling interval as phases and relating the outputs its state and the input symbol. Because CPM can be defined as a finite state machine, the modulation process should be followed on the phase trellis as shown in Figure 2.26. The phase continuity of CPM provides high spectral efficiency. This efficiency can be increased by selecting a better phase modulation function. In Figure 2.27, a sample phase modulation function is given.

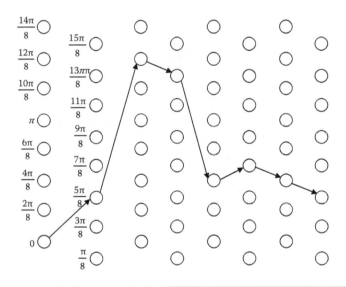

**Figure 2.26  Phase trellis of continuous-phase modulation (CPM).**

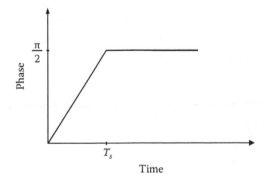

**Figure 2.27  A phase modulation function.**

A continuous phase modulated signal with constant envelope is in the form of

$$s(t,d) = \sqrt{\frac{2E_s}{T}}\cos(2\pi f_c t + \ (t,d) + \ _0)$$

(2.23)

Here, $E_s$ is the symbol energy, $T$ is the symbol duration, $f_c$ is carrier frequency, and $d$ is the input symbols. Information is carried by $(t,d)$, and its phase, during the interval of the symbol $n$th $nT \leq t \leq (n+1)T$, is calculated by the following formula:

$$(t,d) = 2\pi h \sum_{i<n} d_i q \left( t - iT \right)$$

(2.24)

This phase is dependent on the shape of the phase response function, $q(t)$. The phase response of each symbol affects the phase of all signals.

In Figure 2.28, a block diagram of CPM is shown. The definition of CPM is given by Sundberg [8]. Generally, a CPM system can be defined by the instant frequency pulse $g(t)$ which is the derivation of $q(t)$.

$g(t)$ frequency pulse, has nonzero value between $0 \leq t \leq LT$ interval where $T$ is symbol duration, $L$ is the positive constant, and also its value is

$$q(t) = \int_{-\infty}^{\infty} g(t)dt = 1/2$$

$\gamma_i \in \{\pm1, \pm3, \square, \pm(M-1)\}$ is M array information symbols. $h$ is modulation index that affects the phase of $(t,d)$. In general, the modulation provides better error performance for high $h$ values but it needs more bandwidth. The maximum phase difference is $(M-1)\pi h$. Bit energy is computed as $E_b = E_s/\log_2 M$ and bit transfer rate is $T_b = T/\log_2 M$. Various CPM systems can be defined using different $g(t)$, $h$, and $M$.

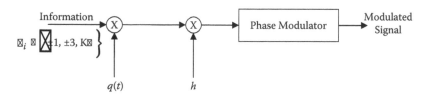

**Figure 2.28  Continuous phase modulator.**

Continuous phase of the signal can be achieved by adding extra memory to the modulation system. Even if the data systems are uncorrelated, memory can be added to the CPM signals by choosing $L > 1$. This modulation scheme is called partial response modulation. In full response modulation $L \leq 1$ is chosen. CPM signals have various trellis structures according to $h$ values. For $h = J/p$ ($J$ and $p$ are positive integers) and during the symbol period of $nT \leq t \leq (n + 1)T$ phase, $(t,d)$ is given as

$$(t,d) = 2\pi h \sum_{i=n-L+1}^{n} d_i q(t - iT) + \theta_n \tag{2.25}$$

and

$$\theta_n = \pi h \sum_{i=-\infty}^{n-L} d_i (\mathrm{mod}\, 2\pi) \tag{2.26}$$

$\theta_n$ is the accumulated phase until the $n$th signaling interval and has $2p$ different values. Therefore, it depends on the data, which remains part of the summation, $d_{n-1}, \ldots, d_{n-L+1}$. CPFSK for $L = 1$ and other full response CPM methods have a number of $2p$ states.

## 2.8.1 Decomposition of CPM

In CPM systems, carrying information continuously by the phase function, and the ability to show the symbols by the phase state trellis including all phase states indicate the presence of a memory structure like convolutional encoder.

Rimoldi [17] shows that a CPM system can be modeled as a continuous phase encoder (CPE) (Figure 2.29) followed by a memoryles modulator (MM). There are two advantages of this decomposition; the first is coding becomes independent from the modulation, and the second is isolating the MM form coding results in modeling the modulator, channel, and demodulator as a separated memoryless channel. A CPE stimulates various CPM and decoding structures for different coding designs. If a CPE is a time invariant linear system, it can be assumed as an encoder. Concatenation of an outer encoder and a CPE can be utilized as a single convolutional encoder, which is investigated in [18,19].

Any CPM system can be decomposed into a CPE and an MM. CPE is a time invariant linear consecutive circuit including modulo-$p$ summation operators and number of $L$ delay elements.

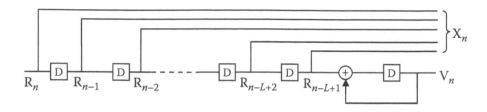

**Figure 2.29   Block diagram of a CPE.**

## 2.8.2  *Continuous Phase Frequency Shift Keying*

The instantaneous frequency pulse response function, $f(t)$, of CPM has a rectangular pulse during time $T$, as shown in Figure 2.30a, and thus its phase response is a trapezoidal function, which means that this modulation is continuous phase frequency shift keying, as given in Figure 2.30b. CPFSK is a continuous phased form of the FSK. However, it not only provides more narrow frequency spectrum but also needs only one oscillator in its modulator. These properties make the CPFSK superior to FSK. Actually to obtain the FSK signals, a number of symbols oscillators should be utilized and oscillators must be switched according to the input bit stream. In practice, this switching cannot be realized ideally and thus some discontinuity occurs in the phase. Therefore the spectrum of the FSK is wider than CPFSK. Using one oscillator in CPFSK prevents phase discontinuities.

The CPFSK signals in the interval $nt \le t \le (n + 1)T$ is

$$s(t,\mathbf{d}) = \sqrt{\frac{2E}{T}} \cos\left[ 2\pi f_c t + 2\pi h \sum_{k=-\infty}^{\infty} d_k \frac{(t - kT)}{2T} \right]$$

$$= \sqrt{\frac{2E}{T}} \cos\left[ 2\pi f_c t + \pi h d_n \frac{(t - nT)}{T} + \varphi_n \right]$$

(2.27)

Here,

$$\varphi_n = \left[ \pi h \sum_{k=-\infty}^{n-1} d_k \right]_{\text{mod } 2\pi}$$

(2.28)

and is the memory that indicates the phase value by summing the phases until $(n-1)T$. If $J$ is even, $\varphi_n \in \{0, \pi J / P, 2\pi J / P, ..., (P-1)\pi J / P\}$. The CPFSK signals cannot be shown in the complex space, such as can the

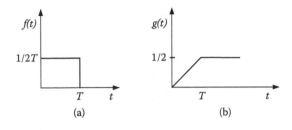

**Figure 2.30** (a) The function *f(t)* of CPFSK; (b) the function *g(t)*, the integral of *f(t)*.

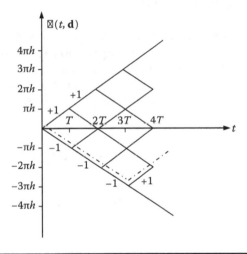

**Figure 2.31** Phase tree of 2-CPFSK.

PSK or QAM, because CPFSK signals can be represented using more than three dimensions. Therefore, as an alternate way to symbolize the CPFSK signals, the phase term $(t,\mathbf{d})$ in Equation (2.27) is denoted using a phase tree as a function of time. In Figure 2.31, the phase tree of 2CPFSK is given employing the starting phase and time as zero. In each modulation interval $T$, for $d_n = (\gamma_n + (M-1))/2$ where $\gamma_n$ is the $n$th information symbol and $\gamma_n = \pm 1$ for 2CPFSK.

In this figure, the input data is assumed as $\gamma = \{-1, +1, -1, -1, +1\}$, and its phase change is demonstrated by dashed lines.

Instead of an enlarging phase tree by time, the phase definition in physical world, the physical phase, can be given between 0 and $2\pi$ as

$$(t,\mathbf{d}) = \left[ (t,\mathbf{d}) \right]_{\mathrm{mod}\, 2\pi} \tag{2.29}$$

Only in this case, the phase trellis can be drawn. The physical phase trellis of the CPFSK signal is shown in Figure 2.32 for $M = 2$ and $h = 1/4$.

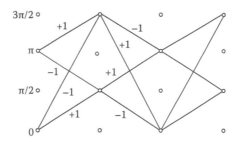

**Figure 2.32  Phase trellis of 2-CPFSK.**

The phase trellis structure varies with time. This makes the decoding process difficult when a channel coding is applied. Therefore, to facilitate encoding and decoding computations, it is better to use a trellis that does not change with time. This sort of CPFSK that does not change with time was introduced by Rimoldi [17,32,33]. Equation (2.27) can be rewritten as

$$s(t,\mathbf{d}) = \sqrt{\frac{2E}{T}} \cos[2\pi f_1 t + \Psi(t,\mathbf{d})] \quad t \geq 0 \tag{2.30}$$

where $f_1 = f_c - b(M-1)/2T$ is a modified carrier frequency and the phase term $\Psi(t,\mathbf{d})$ is

$$\Psi(t,\mathbf{d}) = (t,\mathbf{d}) + \frac{\pi b(M-1)t}{T} \tag{2.31}$$

The physical phase varies between 0 and $2\pi$ so, $\Psi(t,\mathbf{d})$ is

$$\Psi(t,\mathbf{d}) = [\Psi(t,\mathbf{d})]_{\text{mod } 2\pi} \tag{2.32}$$

If we arrange Equation (2.30), the CPFSK signal formula is obtained as

$$s(t,\mathbf{d}) = \sqrt{\frac{2E}{T}} \cos\left[2\pi f_1 t + 2\pi b d_n \frac{(t-nT)}{T} + \theta_n\right] \tag{2.33}$$

$\theta_n$ is the beginning phase angle at time $t = nT$; in other words, it is the memory that provides the phase continuity of the modulation. Moreover,

$$\theta_{n+1} = (\theta_n + 2\pi b d_n)_{\text{mod } 2\pi} \tag{2.34}$$

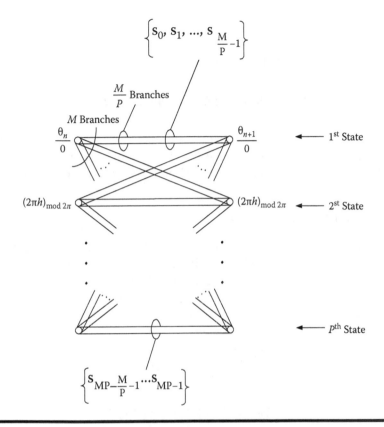

**Figure 2.33** **Time-invariant physical phase trellis of MCPFSK for *h* = *J/P*.**

where $\theta_{n+1}$ and $\theta_n$ take values from the same set. Therefore, $\theta_n \in \{0, (2\pi h)_{\text{mod } 2\pi}, (4\pi h)_{\text{mod } 2\pi}, \ldots\}$ can get $P$ number of different values. A time-invariant $P$ state trellis diagram is shown in Figure 2.33 considering with the phase, $\theta_n$, considered as the state [19,20,29,33]. Furthermore, $s(t, \mathbf{d})$ is the output signal related to input value, $d_n$, and beginning phase, $\theta_n$.

There are number of $M$ output branches from each state. $s_i$ is the $i$th $s(t, \mathbf{d})$ related to $d_n$ and $\theta_n$ values ($i = 0, 1, \ldots, MP - 1$).

The free square Euclidian distance between any two $d$ and $d'$ symbols for MCPFSK and $h = J/P$ can be given as [16,29]

$$
d^2(d, d') = \begin{cases} 2E\left[1 - \dfrac{\sin \Delta_{\Psi_{n+1}} - \sin \Delta_{\Psi_n}}{\Delta_{\Psi_{n+1}} - \Delta_{\Psi_n}}\right], & \Delta_{\Psi_{n+1}} \quad \Delta_{\Psi_n} \\[2em] 2E\left[1 - \cos \Delta_{\Psi_n}\right], & \Delta_{\Psi_n} \end{cases} \tag{2.35}
$$

where

$$\Delta_{\Psi_n} = \Psi(nT,d) - \Psi(nT,d')$$

$$\Delta_{\Psi_{n+1}} = \Psi((n+1)T,d) - \Psi((n+1)T,d')$$

(2.36)

Equation (2.36) indicates the phase differences at time $nT$ and $(n + 1)T$, respectively. In CPFSK, the bandwidth of the signal gets narrow while the modulation index $h$ becomes small. Therefore, $J = 1$ is considered.

### 2.8.2.1 2CPFSK for h = 1/2

The fact that $h = 1/2$ is considered means there are two different phase values in this modulation scheme. These two values can be stored in one memory. The phase trellis and signal set of this modulation scheme are shown in Figure 2.34 [24,29,34]. There are a total of four signals, because $M = 2$ and $P = 2$. 2CPFSK signal formula for $h=1/2$ is

$$s(t,\mathbf{d}) = \sqrt{\frac{2E_s}{T}} \cos\left[ 2\pi f_1 t + \pi d_n \frac{t-nT}{T} + \theta_n \right]$$

(2.37)

$$\gamma_n \in \{-1,+1\} \Rightarrow d_n \in \{0,1\}$$

$$\theta_n \in \{0,\pi\}, \quad R_s = 1 \text{ bit/symbol}, \quad d_{free}^2 = 4E_s$$

In Figure 2.34 and Table 2.5, phase transitions are given where $\theta_n$ and $\theta_{n+1}$ are the beginning and the ending phases of the signal. This phase value is

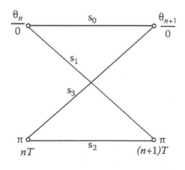

**Figure 2.34** Phase trellis and signal set of 2CPFSK for *h* = 1/2.

**Table 2.8   Phase Transitions and Signal Set of 2CPFSK for $h = 1/2$**

| $\theta_n$ | $d_n$ | $\Psi(t,\mathbf{d})$ | $\Psi(nT,\mathbf{d})$ | $\Psi((n+1)T,\mathbf{d})$ | $s_i$ | $\theta_{n+1}$ |
|---|---|---|---|---|---|---|
| 0 | 0 | 0 | 0 | 0 | $s_0$ | 0 |
|   | 1 | $\pi\dfrac{(t-nT)}{T}$ | 0 | $\pi$ | $s_1$ | $\pi$ |
| $\pi$ | 0 | $\pi$ | $\pi$ | $\pi$ | $s_2 = -s_0$ | $\pi$ |
|   | 1 | $\pi\dfrac{(t-nT)}{T}+\pi$ | $\pi$ | $2\pi$ | $s_3 = -s_1$ | 0 |

Note: $\theta_n$ = beginning angle, $d_n$ = input value of CPE, $\psi(t,\mathbf{d})$ = frequency shifting value, $\psi(nT,\mathbf{d})$ = instant phase value, $\psi((n + 1)T,\mathbf{d})$ = ending real phase, $s_i$ = signal, $\theta_{n+1}$ = ending physical phase.

utilized for the next signaling interval using one memory unit. The memory values of 0 and 1 represent the phase angles of 0 and $\pi$, respectively.

In Table 2.8, $\theta_n$ is the beginning angle, $d_n$ is the input value of CPE, $\psi(t, \mathbf{d})$ is the frequency shifting value, $\psi(nT, \mathbf{d})$ is the instant phase value, $\psi((n + 1)T, \mathbf{d})$ is the ending real phase, $s_i$ is the signal, and $\theta_{n+1}$ is the ending physical phase.

### 2.8.2.2   4CPFSK for $h = 1/2$

The fact that $h = 1/2$ is considered, means there are two different phase values in this modulation scheme. These two values can be stored in one memory. The phase trellis and signal set of this modulation scheme are shown in Figure 2.35 [24,29,34]. There are a total of eight signals, because $M = 4$ and $P = 2$. The 4CPFSK signal formula for $h = 1/2$ is

$$s(t,\mathbf{d}) = \sqrt{\frac{2E_s}{T}}\cos\left[2\pi f_1 t + \pi d_n \frac{t-nT}{T} + \theta_n\right] \tag{2.38}$$

$$\gamma_n \in \{-3,-1,+1,+3\} \Rightarrow d_n \in \{0,1,2,3\}$$

$$\theta_n \in \{0,\pi\}, \; R_s{=}2 \text{ bits/symbol}, \; d_{free}^2 = 4E_s$$

In Figure 2.35 phase transitions are given in which $\theta_n$ and $\theta_{n+1}$ are the beginning and the ending phases of the signal. This phase value is utilized for the next signaling interval using one memory unit. The memory values of 0 and 1 represent the phase angles of 0 and $\pi$, respectively.

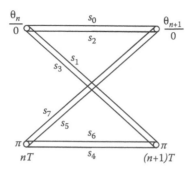

**Figure 2.35 Phase trellis and signal set of 4CPFSK for $h = 1/2$.**

In Table 2.9, $\theta_n$ is the beginning angle, $d_n$ is the input value of CPE, $\psi(t, \mathbf{d})$ is the frequency shifting value, $\psi(nT, \mathbf{d})$ is the instant phase value, $\psi((n + 1)T, \mathbf{d})$ is the ending real phase, $s_i$ is the signal, and $\theta_{n+1}$ is the ending physical phase.

**Table 2.9 Phase Transitions and Signal Set of 4-CPFSK for $h = 1/2$**

| $\theta_n$ | $d_n$ | $\Psi(t,\mathbf{d})$ | $\Psi(nT,\mathbf{d})$ | $\Psi((n+1)T,\mathbf{d})$ | $s_i$ | $\theta_{n+1}$ |
|---|---|---|---|---|---|---|
| 0 | 0 | $0$ | $0$ | $0$ | $s_0$ | $0$ |
| | 1 | $\pi\dfrac{(t-nT)}{T}$ | $0$ | $\pi$ | $s_1$ | $\pi$ |
| | 2 | $2\pi\dfrac{(t-nT)}{T}$ | $0$ | $2\pi$ | $s_2$ | $0$ |
| | 3 | $3\pi\dfrac{(t-nT)}{T}$ | $0$ | $3\pi$ | $s_3$ | $\pi$ |
| $\pi$ | 0 | $\pi$ | $\pi$ | $\pi$ | $s_4 = -s_0$ | $\pi$ |
| | 1 | $\pi\dfrac{(t-nT)}{T}+\pi$ | $\pi$ | $2\pi$ | $s_5 = -s_1$ | $0$ |
| | 2 | $2\pi\dfrac{(t-nT)}{T}+\pi$ | $\pi$ | $3\pi$ | $s_6 = -s_2$ | $\pi$ |
| | 3 | $3\pi\dfrac{(t-nT)}{T}+\pi$ | $\pi$ | $4\pi$ | $s_7 = -s_3$ | $0$ |

*Note:* $\theta_n$ = beginning angle, $d_n$ = input value of CPE, $\psi(t,\mathbf{d})$ = frequency shifting value, $\psi(nT,\mathbf{d})$ = instant phase value, $\psi((n + 1)T, \mathbf{d})$ = ending real phase, $s_i$ = signal, and $\theta_{n+1}$ = ending physical phase.

### 2.8.2.3   8CPFSK for h = 1/2

The fact that $h = 1/2$ is considered means there are two different phase values in this modulation scheme. These two values can be stored in one memory. The phase trellis and signal set of this modulation scheme are shown in Figure 2.36. There are a total of 16 signals, because $M = 8$ and $P = 2$. The 8-CPFSK signal formula for $h = 1/2$ is

$$s(t,\mathbf{d}) = \sqrt{\frac{2E_s}{T}} \cos\left[ 2\pi f_1 t + \pi d_n \frac{t-nT}{T} + \theta_n \right] \tag{2.39}$$

$$\gamma_n \in \{-7,-5-3,-1,+1,+3,+5,+7\} \quad \Rightarrow \quad d_n \in \{0,1,2,3,4,5,6,7\}$$

$$\theta_n \in \{0,\pi\}$$

$$R_s = 3 \text{ bits/symbol}, \ d_{free}^2 = 2E_s$$

In Figure 2.36 phase transitions are given where $\theta_n$ and $\theta_{n+1}$ are the beginning and the ending phases of the signal. This phase value is utilized for the next signaling interval using one memory unit. The memory values of 0 and 1 represent the phase angles of 0 and $\pi$, respectively.

In Table 2.10, $\theta_n$ is the beginning angle, $d_n$ is the input value of CPE, $\psi(t, \mathbf{d})$ is the frequency shifting value, $\psi(nT, \mathbf{d})$ is the instant phase value, $\psi((n + 1)T, \mathbf{d})$ is the ending real phase, $s_i$ is the signal, and $\theta_{n+1}$ is the ending physical phase.

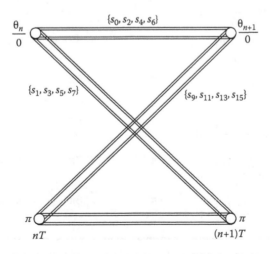

**Figure 2.36   Phase trellis and signal set of 8-CPFSK for $h = 1/2$.**

**Table 2.10    Phase Transitions and Signal Set of 8-CPFSK for $h = 1/2$**

| $\theta_n$ | $d_n$ | $\Psi(t,\mathbf{d})$ | $\Psi(nT,\mathbf{d})$ | $\Psi((n+1)T,\mathbf{d})$ | $s_i$ | $\theta_{n+1}$ |
|---|---|---|---|---|---|---|
| 0 | 0 | $0$ | 0 | 0 | $s_0$ | 0 |
|  | 1 | $\pi\dfrac{(t-nT)}{T}$ | 0 | $\pi$ | $s_1$ | $\pi$ |
|  | 2 | $2\pi\dfrac{(t-nT)}{T}$ | 0 | $2\pi$ | $s_2$ | 0 |
|  | 3 | $3\pi\dfrac{(t-nT)}{T}$ | 0 | $3\pi$ | $s_3$ | $\pi$ |
|  | 4 | $4\pi\dfrac{(t-nT)}{T}$ | 0 | $4\pi$ | $s_4$ | 0 |
|  | 5 | $5\pi\dfrac{(t-nT)}{T}$ | 0 | $5\pi$ | $s_5$ | $\pi$ |
|  | 6 | $6\pi\dfrac{(t-nT)}{T}$ | 0 | $6\pi$ | $s_6$ | 0 |
|  | 7 | $7\pi\dfrac{(t-nT)}{T}$ | 0 | $7\pi$ | $s_7$ | $\pi$ |
| $\pi$ | 0 | $\pi$ | $\pi$ | $\pi$ | $s_8 = -s_0$ | $\pi$ |
|  | 1 | $\pi\dfrac{(t-nT)}{T}+\pi$ | $\pi$ | $2\pi$ | $s_9 = -s_1$ | 0 |
|  | 2 | $2\pi\dfrac{(t-nT)}{T}+\pi$ | $\pi$ | $3\pi$ | $s_{10} = -s_2$ | $\pi$ |
|  | 3 | $3\pi\dfrac{(t-nT)}{T}+\pi$ | $\pi$ | $4\pi$ | $s_{11} = -s_3$ | 0 |
|  | 4 | $4\pi\dfrac{(t-nT)}{T}+\pi$ | $\pi$ | $5\pi$ | $s_{12} = -s_4$ | $\pi$ |
|  | 5 | $5\pi\dfrac{(t-nT)}{T}+\pi$ | $\pi$ | $6\pi$ | $s_{13} = -s_5$ | 0 |
|  | 6 | $6\pi\dfrac{(t-nT)}{T}+\pi$ | $\pi$ | $7\pi$ | $s_{14} = -s_6$ | $\pi$ |
|  | 7 | $7\pi\dfrac{(t-nT)}{T}+\pi$ | $\pi$ | $8\pi$ | $s_{15} = -s_7$ | 0 |

*Note:* $\theta_n$ = beginning angle, $d_n$ = input value of CPE, $\psi(t,\mathbf{d})$ = frequency shifting value, $\psi(nT, \mathbf{d})$ = instant phase value, $\psi((n + 1)T, \mathbf{d})$ = ending real phase, $s_i$ = signal, $\theta_{n+1}$ = ending physical phase.

### 2.8.2.4  16-CPFSK for h=1/2

The fact that $h = 1/2$ is considered means there are two different phase values in this modulation scheme. These two values can be stored in one memory. The phase trellis and signal set of this modulation scheme are shown in Figure 2.37. There is a total of 32 signals, because $M = 16$ and $P = 2$. 16CPFSK signal formula for $h = 1/2$ is

$$s(t,\mathbf{d}) = \sqrt{\frac{2E_s}{T}}\cos\left[2\pi f_1 t + \pi d_n \frac{t-nT}{T} + \theta_n\right] \qquad (2.40)$$

$$\gamma_n \in \{-15,-13,-11,-9-7,-5-3,-1,+1,+3,+5,+7,+9,+11,+13,+15\}$$

$$\Rightarrow d_n \in \{0,1,2,3,4,5,6,7,8,9,10,11,12,13,14,15\}$$

$$\theta_n \in \{0,\pi\}, \; R_s = 4 \text{ bits/symbol}, \; d_{free}^2 = 2E_s$$

In Figure 2.37 phase transitions are given where $\theta_n$ and $\theta_{n+1}$ are the beginning and the ending phases of the signal. This phase value is utilized for the next signaling interval using one memory unit. The memory values of 0 and 1 represent the phase angles of 0 and $\pi$, respectively.

In Table 2.11, $\theta_n$ is the beginning angle, $\beta_n$ is the input value of CPE, $\psi(t, \beta)$ is the frequency shifting value, $\psi(nT, \beta)$ is the instant phase value, $\psi((n + 1)T, \beta)$ is the ending real phase, $s_i$ is the signal, and $\theta_{n+1}$ is the ending physical phase.

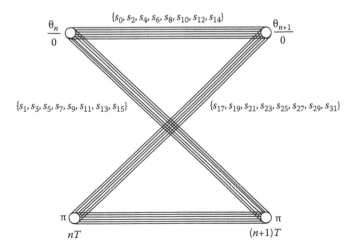

**Figure 2.37    Phase trellis and signal set of 16-CPFSK for $h = 1/2$.**

**Table 2.11   Phase Transitions and Signal Set of 16-CPFSK for $h = 1/2$**

| $\theta_n$ | $d_n$ | $\Psi(t,\mathbf{d})$ | $\Psi(nT,\mathbf{d})$ | $\Psi((n+1)T,\mathbf{d})$ | $s_i$ | $\theta_{n+1}$ |
|---|---|---|---|---|---|---|
| 0 | 0 | $0$ | 0 | 0 | $s_0$ | 0 |
| | 1 | $\pi\dfrac{(t-nT)}{T}$ | 0 | $\pi$ | $s_1$ | $\pi$ |
| | 2 | $2\pi\dfrac{(t-nT)}{T}$ | 0 | $2\pi$ | $s_2$ | 0 |
| | 3 | $3\pi\dfrac{(t-nT)}{T}$ | 0 | $3\pi$ | $s_3$ | $\pi$ |
| | 4 | $4\pi\dfrac{(t-nT)}{T}$ | 0 | $4\pi$ | $s_4$ | 0 |
| | 5 | $5\pi\dfrac{(t-nT)}{T}$ | 0 | $5\pi$ | $s_5$ | $\pi$ |
| | 6 | $6\pi\dfrac{(t-nT)}{T}$ | 0 | $6\pi$ | $s_6$ | 0 |
| | 7 | $7\pi\dfrac{(t-nT)}{T}$ | 0 | $7\pi$ | $s_7$ | $\pi$ |
| | 8 | $8\pi\dfrac{(t-nT)}{T}$ | 0 | $8\pi$ | $s_8$ | 0 |
| | 9 | $9\pi\dfrac{(t-nT)}{T}$ | 0 | $9\pi$ | $s_9$ | $\pi$ |
| | 10 | $10\pi\dfrac{(t-nT)}{T}$ | 0 | $10\pi$ | $s_{10}$ | 0 |
| | 11 | $11\pi\dfrac{(t-nT)}{T}$ | 0 | $11\pi$ | $s_{11}$ | $\pi$ |
| | 12 | $12\pi\dfrac{(t-nT)}{T}$ | 0 | $12\pi$ | $s_{12}$ | 0 |
| | 13 | $13\pi\dfrac{(t-nT)}{T}$ | 0 | $13\pi$ | $s_{13}$ | $\pi$ |
| | 14 | $14\pi\dfrac{(t-nT)}{T}$ | 0 | $14\pi$ | $s_{14}$ | 0 |
| | 15 | $15\pi\dfrac{(t-nT)}{T}$ | 0 | $15\pi$ | $s_{15}$ | $\pi$ |

| $\theta_n$ | | $\psi(t,\mathbf{d})$ | $\psi(nT,\mathbf{d})$ | $\psi((n+1)T,\mathbf{d})$ | $s_i$ | $\theta_{n+1}$ |
|---|---|---|---|---|---|---|
| $\pi$ | 0 | $\pi$ | $\pi$ | $\pi$ | $s_{16}=-s_0$ | $\pi$ |
| | 1 | $\pi\dfrac{(t-nT)}{T}+\pi$ | $\pi$ | $2\pi$ | $s_{17}=-s_1$ | 0 |
| | 2 | $2\pi\dfrac{(t-nT)}{T}+\pi$ | $\pi$ | $3\pi$ | $s_{18}=-s_2$ | $\pi$ |
| | 3 | $3\pi\dfrac{(t-nT)}{T}+\pi$ | $\pi$ | $4\pi$ | $s_{19}=-s_3$ | 0 |
| | 4 | $4\pi\dfrac{(t-nT)}{T}+\pi$ | $\pi$ | $5\pi$ | $s_{20}=-s_4$ | $\pi$ |
| | 5 | $5\pi\dfrac{(t-nT)}{T}+\pi$ | $\pi$ | $6\pi$ | $s_{21}=-s_5$ | 0 |
| | 6 | $6\pi\dfrac{(t-nT)}{T}+\pi$ | $\pi$ | $7\pi$ | $s_{22}=-s_6$ | $\pi$ |
| | 7 | $7\pi\dfrac{(t-nT)}{T}+\pi$ | $\pi$ | $8\pi$ | $s_{23}=-s_7$ | 0 |
| | 8 | $8\pi\dfrac{(t-nT)}{T}+\pi$ | $\pi$ | $9\pi$ | $s_{24}=-s_8$ | $\pi$ |
| | 9 | $9\pi\dfrac{(t-nT)}{T}+\pi$ | $\pi$ | $10\pi$ | $s_{25}=-s_9$ | 0 |
| | 10 | $10\pi\dfrac{(t-nT)}{T}+\pi$ | $\pi$ | $11\pi$ | $s_{26}=-s_{10}$ | $\pi$ |
| | 11 | $11\pi\dfrac{(t-nT)}{T}+\pi$ | $\pi$ | $12\pi$ | $s_{27}=-s_{11}$ | 0 |
| | 12 | $12\pi\dfrac{(t-nT)}{T}+\pi$ | $\pi$ | $13\pi$ | $s_{28}=-s_{12}$ | $\pi$ |
| | 13 | $13\pi\dfrac{(t-nT)}{T}+\pi$ | $\pi$ | $14\pi$ | $s_{29}=-s_{13}$ | 0 |
| | 14 | $14\pi\dfrac{(t-nT)}{T}+\pi$ | $\pi$ | $15\pi$ | $s_{30}=-s_{14}$ | $\pi$ |
| | 15 | $15\pi\dfrac{(t-nT)}{T}+\pi$ | $\pi$ | $16\pi$ | $s_{31}=-s_{15}$ | 0 |

*Note:* $\theta_n$ = beginning angle, $d_n$ = input value of CPE, $\psi(t,\mathbf{d})$ = frequency shifting value, $\psi(nT,\mathbf{d})$ = instant phase value, $\psi((n+1)T,\mathbf{d})$ = ending real phase, $s_i$ = signal, $\theta_{n+1}$ = ending physical phase.

### 2.8.2.5 4CPFSK for h = 1/4

The fact that $h = 1/4$ is considered means there are four different phase values in this modulation scheme. These four values can be stored in two memories. The phase trellis and signal set of this modulation scheme are shown in Figure 2.38. There are a total of 16 signals, because $M = 4$ and $P = 4$. The 4-CPFSK signal formula for $h = 1/4$ is,

$$s(t,\mathbf{d}) = \sqrt{\frac{2E_s}{T}} \cos\left[ 2\pi f_1 t + \frac{\pi}{2} d_n \frac{t - nT}{T} + \theta_n \right] \tag{2.41}$$

$$\gamma_n \in \{-3,-1,+1,+3\} \Rightarrow d_n \in \{0,1,2,3\}$$

$$\theta_n \in \{0, \pi/2, \pi, 3\pi/2\}, \quad R_s = 2 \text{ bits/symbol}, \quad d_{free}^2 = 1.44 E_s$$

In Figure 2.38, phase transitions are given where $\theta_n$ and $\theta_{n+1}$ are the beginning and ending phases of the signal. This phase value is utilized for the next signaling interval using two memory units. The memory values of 00, 01, 10, and 11 represent the phase angles of 0, $\pi/2$, $\pi$, and $3\pi/2$, respectively.

In Table 2.12, $\theta_n$ is the beginning angle, $d_n$ is the input value of CPE, $\psi(t, \mathbf{d})$ is the frequency shifting value, $\psi(nT, \mathbf{d})$ is the instant phase value,

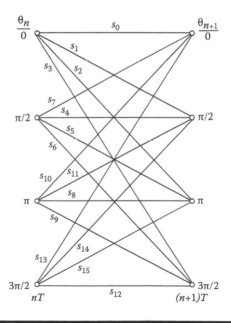

**Figure 2.38  Phase trellis and signal set of 4-CPFSK for h = 1/4.**

$\psi((n + 1)T, \mathbf{d})$ is the ending real phase, $s_i$ is the signal, and $\theta_{n+1}$ is the ending physical phase.

**Table 2.12    Phase Transitions and Signal Set of 4-CPFSK for $h = 1/4$**

| $\theta_n$ | $d_n$ | $\Psi(t,\mathbf{d})$ | $\Psi(nT,\mathbf{d})$ | $\Psi((n+1)T,\mathbf{d})$ | $s_i$ | $\theta_{n+1}$ |
|---|---|---|---|---|---|---|
| $0$ | $0$ | $0$ | $0$ | $0$ | $s_0$ | $0$ |
| | $1$ | $\dfrac{\pi}{2}\dfrac{(t-nT)}{T}$ | $0$ | $\dfrac{\pi}{2}$ | $s_1$ | $\dfrac{\pi}{2}$ |
| | $2$ | $\pi\dfrac{(t-nT)}{T}$ | $0$ | $\pi$ | $s_2$ | $\pi$ |
| | $3$ | $\dfrac{3\pi}{2}\dfrac{(t-nT)}{T}$ | $0$ | $\dfrac{3\pi}{2}$ | $s_3$ | $\dfrac{3\pi}{2}$ |
| $\dfrac{\pi}{2}$ | $0$ | $\dfrac{\pi}{2}$ | $\dfrac{\pi}{2}$ | $\dfrac{\pi}{2}$ | $s_4$ | $\dfrac{\pi}{2}$ |
| | $1$ | $\dfrac{\pi}{2}\dfrac{(t-nT)}{T}+\dfrac{\pi}{2}$ | $\dfrac{\pi}{2}$ | $\pi$ | $s_5$ | $\pi$ |
| | $2$ | $\pi\dfrac{(t-nT)}{T}+\dfrac{\pi}{2}$ | $\dfrac{\pi}{2}$ | $\dfrac{3\pi}{2}$ | $s_6$ | $\dfrac{3\pi}{2}$ |
| | $3$ | $\dfrac{3\pi}{2}\dfrac{(t-nT)}{T}+\dfrac{\pi}{2}$ | $\dfrac{\pi}{2}$ | $2\pi$ | $s_7$ | $0$ |
| $\pi$ | $0$ | $\pi$ | $\pi$ | $\pi$ | $s_8 = -s_0$ | $\pi$ |
| | $1$ | $\dfrac{\pi}{2}\dfrac{(t-nT)}{T}+\pi$ | $\pi$ | $\dfrac{3\pi}{2}$ | $s_9 = -s_1$ | $\dfrac{3\pi}{2}$ |
| | $2$ | $\pi\dfrac{(t-nT)}{T}+\pi$ | $\pi$ | $2\pi$ | $s_{10} = -s_2$ | $0$ |
| | $3$ | $\dfrac{3\pi}{2}\dfrac{(t-nT)}{T}+\pi$ | $\pi$ | $\dfrac{5\pi}{2}$ | $s_{11} = -s_3$ | $\dfrac{\pi}{2}$ |
| $\dfrac{3\pi}{2}$ | $0$ | $\dfrac{3\pi}{2}$ | $\dfrac{3\pi}{2}$ | $\dfrac{3\pi}{2}$ | $s_{12} = -s_4$ | $\dfrac{3\pi}{2}$ |
| | $1$ | $\dfrac{\pi}{2}\dfrac{(t-nT)}{T}+\dfrac{3\pi}{2}$ | $\dfrac{3\pi}{2}$ | $2\pi$ | $s_{13} = -s_5$ | $0$ |
| | $2$ | $\pi\dfrac{(t-nT)}{T}+\dfrac{3\pi}{2}$ | $\dfrac{3\pi}{2}$ | $\dfrac{5\pi}{2}$ | $s_{14} = -s_6$ | $\dfrac{\pi}{2}$ |
| | $3$ | $\dfrac{3\pi}{2}\dfrac{(t-nT)}{T}+\dfrac{3\pi}{2}$ | $\dfrac{3\pi}{2}$ | $3\pi$ | $s_{15} = -s_7$ | $\pi$ |

*Note:* $\theta_n$ = beginning angle, $d_n$ = input value of CPE, $\psi(t,\mathbf{d})$ = frequency shifting value, $\psi(nT,\mathbf{d})$ = instant phase value, $\psi((n + 1)T,\mathbf{d})$ = ending real phase, $s_i$ = signal, $\theta_{n+1}$ = ending physical phase.

# References

[1] Oppenheim, A. V., and R. W. Schafer. *Digital Signal Processing*. Upper Saddle River, NJ: Prentice Hall, 1975.

[2] U•an, O. N., and A. Muhittin Albora. *New Approaches in Signal and Image Processing* (in Turkish). Istanbul: Istanbul University Press, 2003.

[3] U•an, O. N., and O. Osman. *Computer Networks and Communications Techniques* (in Turkish). Istanbul: Istanbul University Press, 2003.

[4] Gose, E., H. Pastaci, O. N. U•an, K. Buyukatak, and O. Osman, Performance of Transmit Diversity-Turbo Trellis Coded Modulation (TD-TTCM) over Genetically Estimated WSSUS MIMO Channels. *Frequenz,* 58:249–255, 2004.

[5] U•an, O. N., and O. Osman. Concatenation of Space-Time Block Codes and Turbo Trellis Coded Modulation (ST-TTCM) Over Rician Fading Channels with Imperfect Phase. *Int. J. Commun. Systems 17*(4):347–361, 2004.

[6] Hekim, Y. Error Performance of Low Density Parity Check Codes. MSc Thesis, Istanbul University, Istanbul, 2005.

[7] Shannon, C. E. Probability of Error for Optimal Codes in Gaussian Channel. *Bell Syst. Tech. J.* 38:611–656, 1959.

[8] Sundberg, C. E. Continuous Phase Modulation. *IEEE Trans. Comm. Mag.* 1:25–38, 1986.

[9] Anderson, J. B. *Digital Phase Modulation*. New York: Plenum 1986.

[10] Anderson, J. B., and D. P. Taylor. A Bandwidth Efficient Class of Signal Space Codes. *IEEE Trans. Inform. Theory* IT-24:703–712, 1978.

[11] Aulin, T., and C. E. Sundberg. Continuous Phase Modulation-Part I: Full Response Signaling. *IEEE Trans. Comm.* COM-29(3):196–209, 1981.

[12] Aulin, T., and C. E. Sundberg. Continuous Phase Modulation-Part II: Partial Response Signaling. *IEEE Trans. Comm.* COM-29(3):210–225, 1981.

[13] Ekanayake, N., M-ary Continuous Phase Frequency Shift Keying with Modulation Index 1/M. *IEEE Proceedings-F,* 131(2):173–178, 1984.

[14] Anderson, J. B., and C. E. Sundberg. Advances in Constant Envelope Coded Modulation. *IEEE Trans. Comm. Mag.*, 36–45, 1991.

[15] Sipser, M., and D. Spielman. Expander Codes. *IEEE Trans. Inform. Theory* 42:1710–1722, 1996.

[16] Lee, E. A., and C. David. *Digital Communication*. New York: Kluwer Academic Publishers, 1994.

[17] Rimoldi, B. E. A Decomposition Approach to CPM. *IEEE Trans. Inform. Theory* 34(2):260–270, 1988.

[18] Pizzi, S. V., and S. G. Wilson. Convolutional Coding Combined with Continuous Phase Modulation. *IEEE Trans. Comm.* COM-33:20–29, 1985.

[19] Lindell, G. Minimum Euclidean Distance for Short Range 1/2 Convolutional Codes and CPFSK Modulation. *IEEE Trans. Inform. Theory* IT-30:509–520, 1984.

[20] Altunbas, I., and U. Aygolu. Design and Performance Analysis of Multilevel Coded M-ary CPFSK. *Proc. Telecommun. Conf. ICT'96,* 269–272, 1996.

[21] Goodman, J. Trends in Cellular and Cordless Communication. *IEEE Comm. Mag.*, 31–40, 1993.

[22] Leib, H., and S. Pasupathy. Error Control Properties of Minimum Shift Keying. *IEEE Comm. Mag.*:52–61, 1993.

[23] Ganesan, A. Capacity Estimation and Code Design Principles for Continuous Phase Modulation (CPM). MSc Thesis, Texas A&M University, Texas, 2003.

[24] Niyazi O. Error Performance of Continuous Phase Source/Channel Coded Systems in Fading Channels. MSc Thesis, Istanbul University, Istanbul, 2002.

[25] U•an, O. N., and O. Osman. *Communication Theory and Engineering Applications* (in Turkish). Ankara, Turkey: Nobel Press, 2006.

[26] U•an, O. N., O. Osman, and A. Muhittin Albora. *Image Processing and Engineering Applications* (in Turkish). Ankara, Turkey: Nobel Press, 2006.

[27] Muhittin Albora, A., O. N. U•an, and O. Osman. *Image Processing in Geophysics Engineering Technical Applications* (in Turkish). Ankara, Turkey: Nobel Press, Ankara, 2006.

[28] U•an, O. N., and O. Osman. *Computer Networks and Network Security* (in Turkish). Ankara, Turkey: Nobel Press, 2006.

[29] Altunbas, I. Multi Level Coding of Continuous Phase Modulation. Ph.D. thesis, Istanbul Technical University, Science and Technology Institute, Istanbul, 1999.

[30] Glover, I. A., and P. M. Grant. *Digital Communications*. New York: Prentice Hall, 1998.

[31] Carlson, A. B. *Communication Systems: An Introduction to Systems and Noise* (3rd edition). New York: McGraw-Hill, 1986.

[32] Rimoldi, B. Design of Coded CPFSK Modulation Systems for Bandwidth and Energy Efficiency. *IEEE Trans. Commun.* 37(9):897–905, 1989.

[33] Altunbas, I., and U. Aygolu, Multilevel Coded CPFSK Systems for AWGN and Fading Channels. *IEEE Trans. Commun.* 48(5):764–773, 2000.

[34] Osman, O. Performance of Turbo Coding. PhD Thesis, Istanbul University, Turkey, 2003.

# Chapter 3

# Multiplexing

In telecommunications and computer networks, multiple analog message signals or digital data streams are combined into one signal to share the expenses of the resources. This process is called multiplexing. In communications, the multiplexed signal is transmitted over a communication channel. In multiplexing processes, a low-level communication channel is divided into various high-level channels to transfer each message signal or data stream. Demultiplexing is the reverse process; it is applied to extract the original channels on the receiver side.

In this chapter, fundamentals of multiple access technologies are explained. These are time division multiple access (TDMA), code division multiple access (CDMA), frequency division multiple access (FDMA), and orthogonal frequency division multiplexing (OFDM) [1–8].

## 3.1 Time Division Multiple Access

The media is used consecutively by the users in TDMA. It is a type of time division multiplexing (TDM), but there are multiple transmitters instead of one transmitter connected to the receiver. A certain timeslot is assigned to each user and this interval is used by only that user during the transmission. A physical channel is defined as a timeslot and the user can access this channel periodically, and no one has the right to use the channel in another user's period, even if that timeslot will not be used. Voice, data, and signaling information are separated into different timeslots. The interval of a timeslot is generally interpreted in milliseconds.

It is very important to avoid overlapping the adjacent timeslots. For example, in large cellular systems, different time delays result in overlapping of the consecutive timeslots. Therefore, perfect synchronization is needed between the transmitters. Interference between the adjacent timeslots should be prevented by putting some guard time on both sides of each timeslot. Synchronization, augmented bandwidth, and power problems are the disadvantages of TDMA. Transmitting and receiving processes have a discontinuous manner [1–5]. In Figure 3.1, an example TDMA signaling is shown.

One of the solutions for the synchronization problem is asynchronous transfer mode (ATM). ATM is one of the most frequently used communication techniques. The key feature of this method is that synchronization between the transmitter and the receiver is not necessary while the communication is going on. In ATM, different transfer rates are possible depending on the communication channel factors. The difference between ATM and TDM is shown in Figure 3.2.

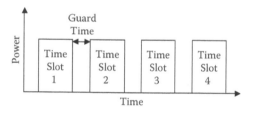

**Figure 3.1 TDMA signaling.**

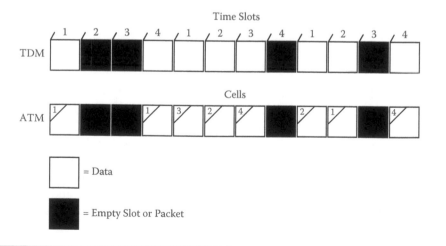

**Figure 3.2 Comparison of TDM and ATM (asynchronous transfer mode).**

The main difference between TDM and TDMA is that the duration of timeslots of TDMA can be updated. However, it is not enough to increase the efficiency by changing this duration. Because the number of users, transferred data size, and type don't vary very quickly. Generally, users do not know if there are other users ready to send data at that instant. If so, the related timeslot would be empty and thus, the efficiency becomes less. A solution is to develop network technologies called ATM. ATM is an asynchronous 53-byte packet transmission [5] and all users have a right to transmit a packet at any time only if the media is empty.

## 3.2 Code Division Multiple Access

CDMA is a technique of multiple access that does not divide the channel into timeslots or frequency bands as in TDMA and FDMA. Instead, CDMA encodes data with a special code (pseudo noise sequences for reverse and Walsh codes for forward channels) associated with each channel and uses the constructive interference properties of the special codes to perform the multiplexing as shown in Figure 3.3. CDMA is a digital technology that is used especially for mobile telephone systems based on IS-95; the IS refers to an interim standard of the Telecommunications Industry Association (TIA). Moreover, CDMA is known as a spread-spectrum technology. The military has used this technology since the 1940s because of its security advantages. On the other hand, the global system for mobile communications (GSM) standard is a specification of an entire network infrastructure. The CDMA interface refers only to the air interface portion of the GSM standard [6].

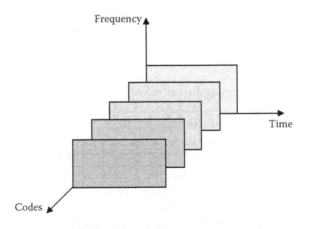

**Figure 3.3  Code division multiple access (CDMA).**

The CDMA technique is used in many digital communication systems assigning the same frequency to all users (Figure 3.3), where each user is assumed to have a different code for the same carrier frequency.

There are three classes of CDMA based on the modulation method: time-hopping CDMA (TH-CDMA), frequency-hopping CDMA (FH-CDMA), and direct-sequence CDMA (DS-CDMA).

In TH-CDMA, first, time intervals are determined by the code assigned to the user. Then, signals are transmitted in rapid burst form at these predetermined intervals. The time axis includes frames in which each frame has $M$ timeslots. The user inserts the data into one of the $M$ timeslots. $M$ timeslots are transmitted based on the code signal assigned to the user. In TH-CDMA, a user transmits all of its data in one timeslot, therefore the frequency increases by a factor of $M$.

In FH-CDMA, the carrier frequency of the modulated signal changes periodically. The carrier frequency is kept constant during time intervals $T$. After each time interval, the carrier jumps to another frequency. The spreading code determines the hopping pattern.

In DS-CDMA, the information signal is modulated by a digital code signal. Generally, the information signal is a digital signal, but it can also be analog. If it is digital the data signal is multiplied by the code signal and the resulting signal modulates the wideband carrier. To achieve higher speeds and support more users compared to the GSM networks based on the TDMA signaling method, wideband CDMA (W-CDMA) is used as a wideband spread-spectrum technique utilizing the DS-CDMA signaling method.

All terminals use the same frequency at the same time and thus, CDMA benefits from the entire bandwidth. Each user has a random unique code and applies exclusive OR (XOR) operation to this code before sending the signal. This signal could be obtained by the receiver who knows this random code. CDMA has a code set that is larger than the frequency set.

### 3.2.1 Wideband CDMA

W-CDMA is the most widespread third-generation (3G) technology. It provides high capacity for voice and data, as well as a high data transfer rate. W-CDMA transmits data on a pair of five MHz-wide radio channels and uses a DS-CDMA transmission technique. W-CDMA has a capacity of up to 2 Mbps for local areas and 384 Kbps for wide areas. W-CDMA is also known as the universal mobile telecommunications system (UMTS), which is one of the 3G mobile phone technologies.

## 3.2.2 CDMA Cellular System

CDMA cellular is an approach that increases the efficiency of CDMA systems by reusing the frequencies, which is the main point of cellular systems. Figure 3.4 shows a model of the CDMA cellular system.

Here, MS, BS, MSC, and BSC represent a mobile station, base station, mobile station controller, and base station controller respectively. CDMA cells can be separated in various numbers of clusters according to the geographical property of the region. The reuse distance is adjusted to prevent interference between base stations. Thus, CDMA code and the frequency range of neighbor, or close, cells are different. To minimize the interference between near cells, various placement configurations are developed as in Figure 3.5.

Handover or handoff is a crucial property in cellular systems. It transfers the connection to other cells without any interruption in connection and without causing the mobile unit to perceive the transition while moving from one cell to another. A cell phone is connected to two or more cells simultaneously during a call, which is a property of the soft handoff (Figure 3.6). It is possible to receive the same signals simultaneously from two or more base stations. If the power of the received signals is nearly the same, the subscriber station receiver can combine these signals in such a way that the bit stream is decoded more reliably than if only one signal is received by the subscriber station. Otherwise, the low-power signal is ignored, and the relatively high-power signal is taken into account by the radio base stations.

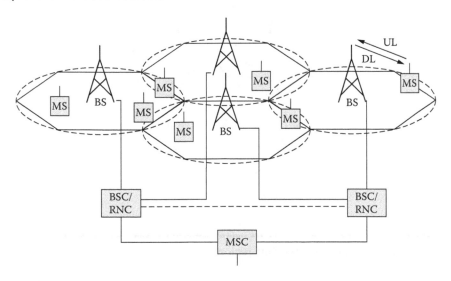

**Figure 3.4   Cellular mobile communication scheme.**

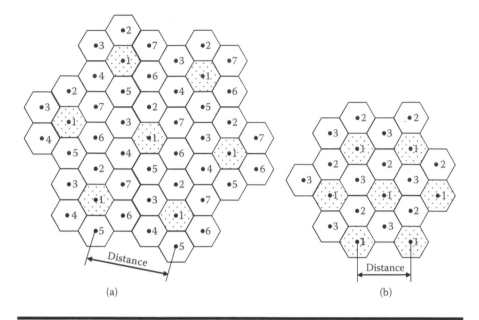

**Figure 3.5** **Reuse of frequencies according to (a) seven cells per cluster (b) three cells per cluster.**

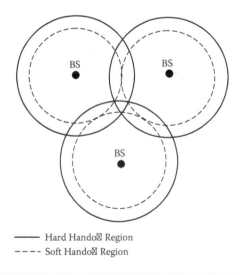

——— Hard Handoff Region
- - - - Soft Handoff Region

**Figure 3.6** **Hard and soft handoff regions.**

In the case of the reverse link, mobile phone to cell site, all the cell site sectors that are actively supporting a call in a soft handover send the bit stream that they receive back to the BSC, along with information about the quality of the received bits. The BSC analylizes the quality of all these bit streams and dynamically chooses the highest quality bit stream.

## 3.3 Frequency Division Multiple Access

In frequency division multiple access (FDMA), total available bandwidth is divided into physical channels that have the same bandwidths as shown in Figure 3.7. Each physical channel or frequency band is assigned to one user. Assigned channels cannot be used by other users during a call. The same channel can be assigned to another user only if the initial call is finished. FDMA is an access technology that shares the radio spectrum among subscribers.

## 3.4 Orthogonal Frequency Division Multiplexing

Orthogonal frequency division multiplexing (OFDM) is a special case of a multicarrier communication system in which a bit stream is sent over a certain number of lower rate subcarriers [7,8]. Here, OFDM can assume either a sort of modulation technique or a multiplexing technique. One of the most important reasons for choosing this method is its power against frequency selective fading and narrowband interference. Fading effect or interference causes an entire communication failure in single-carrier systems, but in multicarrier systems only a few subcarriers are affected. The affected subcarriers can be corrected by using error correction codes.

In frequency division multiplexing, a total signal frequency band is divided into $N$ nonoverlapping channels. Each subchannel is modulated with a separate symbol and a different subcarrier. The easiest way to eliminate the interchannel interference is to avoid spectral overlap of

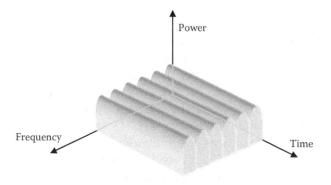

**Figure 3.7 Frequency division multiple access.**

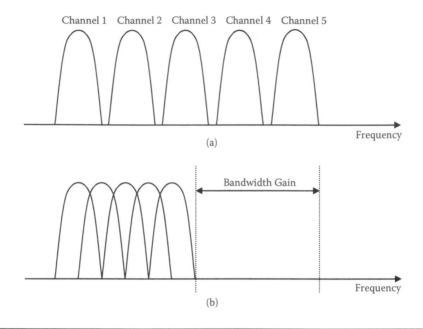

**Figure 3.8** **(a) Conventional multicarrier modulation technique and (b) orthogonal multicarrier modulation technique.**

channels, but results in inefficient use of the overall spectrum. This disadvantage can be eliminated by overlapping the subchannels called OFDM.

In OFDM systems, channel parameters are estimated using the pilot symbol-aided method. OFDM is used in digital video broadcasting—terrestrial (DVB-T) and digital satellite communications channels. In Figure 3.8, conventional nonoverlapping and overlapping multicarrier modulation techniques are shown to emphasize the difference between these two methods. Overlapping the channel frequencies provides a 50 percent gain in bandwidth. However, this could be realized if all the carriers are orthogonal.

Orthogonality signifies that there is a mathematical relationship between the frequencies of the carriers. In the FDM system, conventional filters and demodulators can be used to detect the carriers that are separated in the frequency. To reduce intercarrier interference, guard bands are introduced between the carriers; however, this results in a lowering of the spectrum efficiency.

Although the sidebands of the individual carriers are overlapped in OFDM signals, it is possible to receive the signal without adjacent carrier interface if the carriers are mathematically orthogonal. There are a number of carrier demodulators that multiply each received signal by all carriers then integrate them over a symbol period. If a received signal is multiplied

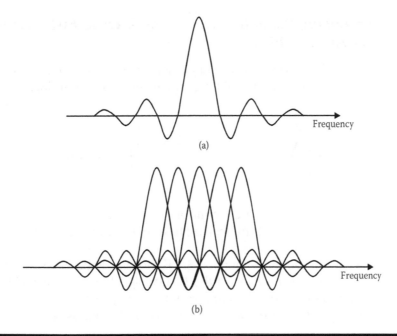

**Figure 3.9** **(a) An OFDM subchannel spectrum and (b) an OFDM signal spectrum.**

by a different carrier, the result of the integration should be zero because of the orthogonality.

In Figure 3.9 (a) and (b), the frequency spectrum of a subchannel and six OFDM signals are shown. In Figure 3.9 (b), there is no interference at the center frequency of each subcarrier because of neighbor channels. Therefore, if discrete Fourier transform (DFT) is used at the receiver, and if the correlation values with the center of frequency of each subcarrier are calculated, the transmitted data can be recovered without interference.

Fast Fourier transformation (FFT) can be achieved by a digital hardware as an application of DFT to dismiss the sets of the subchannels oscillators of FDM.

In recent years, developments in very large scale integration (VLSI) technology made large-scale and high-speed FFT chips commercially affordable. Moreover, the number of computations decreased to $N log N$ from $N^2$ by using FFT technique in both transmitter and receiver sides.

OFDM was used in many military systems in the early years; then it was applied to high-speed modems, digital-mobile communications, and high-density recordings. After the 1990s, OFDM became dominant within wideband communications over mobile radio channels, high-bit rate digital subscriber line (HDSL), asymmetric digital subscriber line (ADSL), very high speed digital subscriber lines (VDSL), digital audio broadcasting (DAB), and high-definition television (HDTV) [8–14].

### 3.4.1 Producing the Subcarrier Using Inverse Fast Fourier Transform (IFFT)

An OFDM signal is formed from subcarriers that are the modulation of PSK or QAM symbols. The OFDM signal can be given as in [8]

$$s(t) = \text{Re}\left\{ \sum_{i=-\frac{N_s}{2}}^{\frac{N_s}{2}-1} x_{i+N_s/2} e^{(j2\pi(f_c - \frac{i+0.5}{T})(t-t_b))} \right\}, \quad t_b \leq t \leq t_b + T \quad (3.1)$$

where $x_i$ is complex PSK or QAM symbol, $N_s$ is the number of subcarriers, $T$ is the symbol duration, $f_c$ is the carrier frequency, and $t_b$ is the beginning time.

The baseband notation of Equation 3.1 is given in Equation 3.2 as in [8]. The real and the imaginary parts of the OFDM signal have to be multiplied by the cosine and sine of the desired carrier frequency to produce the OFDM signal. In Figure 3.10, an OFDM modulator block diagram is shown.

$$s(t) = \sum_{i=-\frac{N_s}{2}}^{\frac{N_s}{2}-1} x_{i+N_s/2} e^{(j2\pi\frac{i}{T}(t-t_b))}, \quad t_b \leq t \leq t_b + T \quad (3.2)$$

In Figure 3.11, an OFDM signal with six subcarriers is depicted. Here, all subcarriers have the same phase and amplitude but both values change according to the outputs of the IFFT.

The baseband representation of an OFDM signal in discrete time is given in Equation 3.3, which is the inverse Fourier transform of $N_s$ PSK or QAM input symbols [4,5]. In practice, this transformation can be implemented by the inverse fast Fourier transform (IFFT).

$$s(n) = \sum_{i=0}^{N_s-1} x_i e^{j2\pi\frac{in}{N}} \quad (3.3)$$

The complexity of IFFT can be reduced by using a radix-4 algorithm. Figure 3.12 shows the four-point radix-4 butterfly IFFT, which is the basis for constructing large IFFT sizes. The output values from $y_0$ to $y_3$ are the transformations of the input values from $x_0$ to $x_3$ using simple additions and phase rotations.

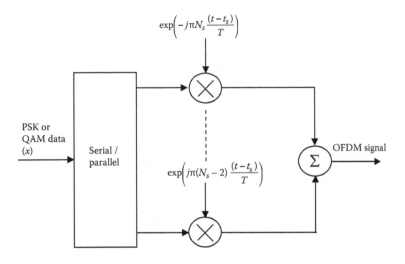

**Figure 3.10  OFDM modulator.**

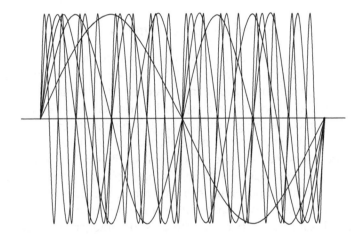

**Figure 3.11  OFDM signal with six subcarriers.**

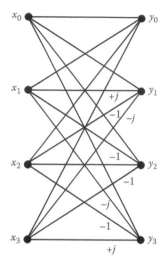

**Figure 3.12  Radix-4 butterfly.**

## 3.4.2  Guard Time and Cyclic Extension

OFDM is one of the most efficient ways to combat multipath delay spread. A guard time is added for each OFDM symbol as a prefix that almost completely eliminates intersymbol interference (ISI). The guard time should be larger than the expected delay spread to prevent interference between consecutive symbols. If the guard time has no signal, intercarrier interference (ICI) will occur. ICI is a type of crosstalk that causes interference between different subcarriers that are not orthogonal. In Figure 3.13, the ICI effect is illustrated with three subcarriers. Here, a subcarrier 1, an expedited subcarrier 2, and a delayed subcarrier 3 are shown. During the demodulation process of the first subcarrier, it will encounter some interference from the second and the third subcarriers because there is no integer number of cycles difference between subcarriers. In the same way, crosstalk will arise from the first and the third subcarriers to the second subcarrier and from the first and the second subcarriers to the third subcarrier for the same reason [7].

In Figure 3.14, OFDM symbols are cyclically extended in the guard time; hence, ICI would be eliminated. This guarantees that delayed or

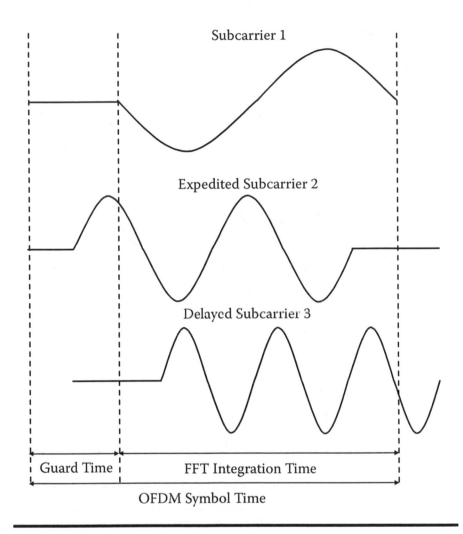

**Figure 3.13   Effect of zero signal in the guard time.**

expedited subcarriers always have an integer number of cycles within the FFT interval. This case is valid only if delay spread is smaller than the guard time. Therefore, if the delays of the multipath signals are smaller than the guard time, there is no ICI at all.

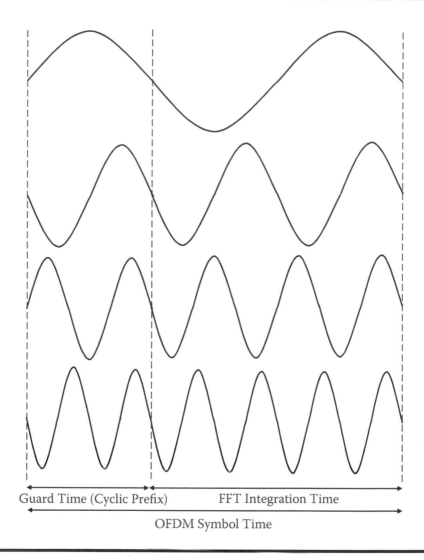

Guard Time (Cyclic Prefix)    FFT Integration Time

OFDM Symbol Time

**Figure 3.14   OFDM signal with cyclic prefix.**

### 3.4.3  Windowing

Modulation causes sharp phase transitions between the symbol transitions. Therefore, phase discontinuity occurs and thus, subcarrier spectrum goes out of the band, and intercarrier interference occurs. To prevent ICI, the spectrum should go down more rapidly. Therefore, windowing is generally applied to each OFDM symbol. Windowing is a process that multiplies the symbols and makes the amplitude go smoothly to zero at the symbol boundaries of an OFDM symbol. The raised cosine function is known as a commonly used window type, which is defined in [8] for OFDM symbols as

$$
w(t) = \begin{cases} 0.5 + 0.5\cos(\pi + t\pi/(\beta T_s)) & 0 \le t \le \beta T_s \\ 1.0 & \beta T_s \le t \le T_s \\ 0.5 + 0.5\cos((t - t_b)\pi/(\beta T_s)) & T_s \le t \le (1+\beta)T_s \end{cases} \tag{3.4}
$$

where $T_s$ is the symbol interval, which is shorter than the total symbol duration because partial overlapping between OFDM symbols is allowed in the roll-off region. Here, symbol duration $T + T_{guard} < T_{pre``x} + T + T_{post``x}$. The multiplier parameter, which is known as windowing, looks like Figure 3.15 [8]. The OFDM symbol signal starting at time $t = t_b = kT_s$ is defined as [8]

$$
s_k(t) = \mathrm{Re}\left\{ w(t - t_b) \sum_{i=-\frac{N_S}{2}}^{\frac{N_S}{2}-1} d_{i+N_S(k+1/2)} e^{(j2\pi(f_c - \frac{i+0.5}{T})(t - t_b - T_{prefix}))} \right\}, \tag{3.5}
$$

$$
t_b \le t \le t_b + T_s(1+\beta)
$$

In wireless communication systems, OFDM signals are generated as follows. First, input bit streams are encoded using channel coding techniques, as explained in Chapter 4, and mapped to PSK or QAM values. Then, the output complex values are put into parallel form and padded with zeros to get a number that is the power of two. These symbols are entered into the IFFT block. The last $T_{pre``x}$ samples of the IFFT output are appended at the beginning of the OFDM symbol and the first $T_{post``x}$ samples are inserted at the end. Before sending the OFDM

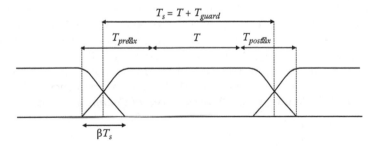

**Figure 3.15   OFDM cyclic extension and windowing.**

symbol, the power of the out-of-band subcarriers is decreased by multiplying a raised cosine window. Finally, the OFDM symbol is appended to the output of the previous OFDM symbol with an overlap region of $\beta T_s$. Here is the roll-off factor of the raised cosine function as shown in Figure 3.15 [8].

# References

[1] Oppenheim, A. V., and R. W. Schafer. *Digital Signal Processing.* Upper Saddle River, NJ: Prentice Hall, 1975.
[2] Glover, I. A., and P. M. Grant. *Digital Communications.* Upper Saddle River, NJ: Prentice Hall, 1998.
[3] Proakis, J.G. *Digital Communications.* New York: McGraw Hill, 2001.
[4] U•an, O. N., and O. Osman. *Computer Networks and Communications Techniques* (in Turkish). Istanbul: Istanbul University Press, 2003.
[5] U•an, O. N., and O. Osman. *Communications Theory and Engineering Applications* (in Turkish). Ankara, Turkey: Nobel Press, 2006.
[6] Viterbi, A. J. *CDMA Principles of Spread Spectrum Communication.* Reading, MA: Addison Wesley, 1997.
[7] Bahai, A. R. S., B. R. Saltzberg, and M. Ergen. *Multi Carrier Digital Communications: Theory and Applications of OFDM.* New York: Springer, 2004.
[8] van Nee, R., and R. Prasad. *OFDM Wireless Multimedia Communications.* New York: Artech House, 2000.
[9] Chow, P. S., J. C. Tu, and J. M. Cioffi. Performance Evaluation of a Multichannel Transceiver System for ADSL and VHDSL Services. *IEEE J. Selected Areas in Comm.* SAC-9(6):909–919, 1991.
[10] Chow, P. S., J. C. Tu, J. M. Cioffi. A Discrete Multitone Transceiver System for HDSL Applications. *IEEE J. Selected Areas in Comm.* SAC-9(6):895–908, 1991.
[11] Paiement, R.V. Evaluation of Single Carrier and Multicarrier Modulation Techniques for Digital ATV Terrestarial Broadcasting. *CRC Report*, No. CRC-RP-004. Ottowa, Canada, 1994.
[12] Sari, H., G. Karma, and I. Jeanclaude. Transmission Techniques for Digital Terrestrial TV Broadcasting. *IEEE Comm. Mag.* 33:100–109, 1995.
[13] Oppenheim, A.V., and R. W. Schaffer. *Discrete-Time Signal Processing.* Englewood Cliffs, NJ: Prentice Hall International, 1989.
[14] Hara, S., M. Mouri, M. Okada, and N. Morinaga N. Transmission Performance Analysis of Multi-Carrier Modulation in Frequency Selective Fast Rayleigh Fading Channel. *Wireless Personal Comm.* 2:335–356, 1996.

# Chapter 4

## Fundamental Coding Techniques

### 4.1 Source Coding

The source entropy is a fundamental property of the source. Source coding does not change the source entropy but usually increases the entropy of the source coded symbols. In this section, the variable length source coding technique, Huffman coding, will be explained.

#### 4.1.1 Huffman Codes

The lengths of the code words used in block codes are always constant. However, if the probabilities of the messages are not the same, efficiency becomes low. To cope, code words with different code lengths are used. Huffman coding is a uniquely and instantaneously decipherable source coding technique. In other words, for a uniquely decipherable coding, there should be a one-to-one correspondence between the messages and the code words. In addition, none of these code words are formed by combinations of two or more other code words. Instantaneously decipherable codes can be sensed when its last bit is received without having to wait to decode the message as in the case of just uniquely decipherable codes. This is possible if the codes are uniquely decipherable and if any code is not the prefix of any other code.

Let the messages (for which we want to find their Huffman codes) be $s_1$, $s_2$, $s_3$, $s_4$, and $s_5$, and their probabilities 4/30, 2/30, 3/30, 1/2, and 1/5, respectively. Huffman codes can be obtained in four steps.

First of all, the messages are arranged according to their probabilities in descending order as shown in Figure 4.1. The orders of the messages with the same probabilities are not important.

| Message | Probability |
|---------|-------------|
| $s_4$ | 1/2 |
| $s_5$ | 1/5 |
| $s_1$ | 4/30 |
| $s_3$ | 3/30 |
| $s_2$ | 2/30 |

**Figure 4.1    First step of Huffman coding.**

The two lowest probabilities in the list are added and the new list is reordered in descending order as shown in Figure 4.2.

| Message | Probability | Probability |
|---------|-------------|-------------|
| $s_4$ | 1/2 | 1/2 |
| $s_5$ | 1/5 | 1/5 |
| $s_1$ | 4/30 | 5/30 |
| $s_3$ | 3/30 | 4/30 |
| $s_2$ | 2/30 | |

$s_2 s_3$

**Figure 4.2    Second step of Huffman coding.**

The addition of the two lowest values and reordering the list continue until two probabilities are left.

Code words are assigned starting from the right-hand side of Figure 4.3. If there is a separation when moving to the left, the assigned bits are shifted one bit to the left and zero is assigned to the upper code while one is assigned to the lower as shown in Figure 4.4. The completed code words are underlined.

Finally, the code words of the corresponding messages are obtained as in Figure 4.5.

The average code length is

$$L = 3 \times 4/30 + 4 \times 2/30 + 4 \times 3/30 + 1 \times 1/2 + 2 \times 1/5$$
$$= 53/30$$
$$= 1.77 \text{ bits}$$

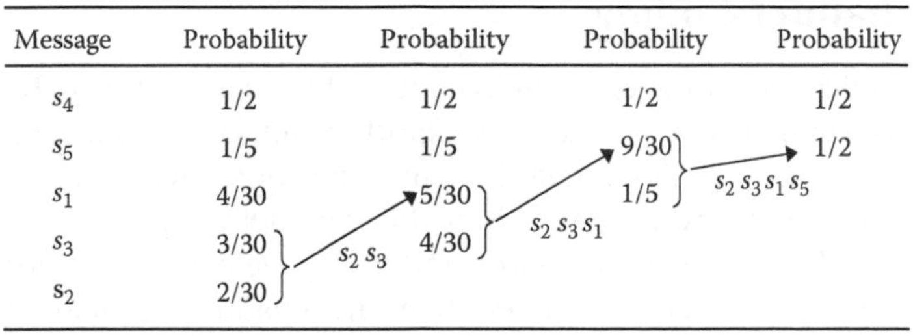

| Message | Probability | Probability | Probability | Probability |
|---------|-------------|-------------|-------------|-------------|
| $s_4$ | 1/2 | 1/2 | 1/2 | 1/2 |
| $s_5$ | 1/5 | 1/5 | 9/30 | 1/2 |
| $s_1$ | 4/30 | 5/30 | 1/5 | $s_2 s_3 s_1 s_5$ |
| $s_3$ | 3/30 | 4/30 | $s_2 s_3 s_1$ | |
| $s_2$ | 2/30 | $s_2 s_3$ | | |

**Figure 4.3   Third step of Huffman coding.**

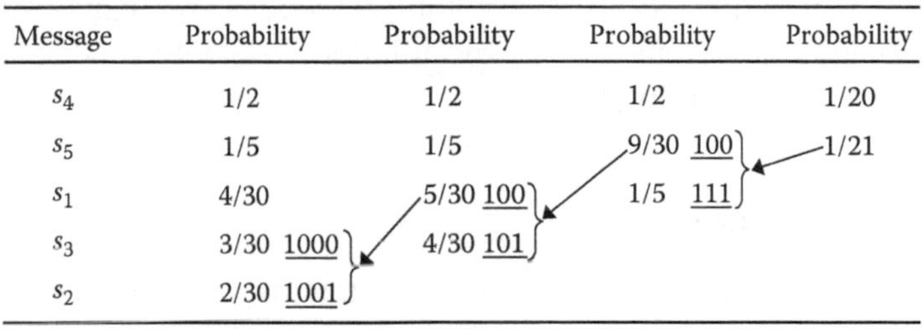

| Message | Probability | Probability | Probability | Probability |
|---------|-------------|-------------|-------------|-------------|
| $s_4$ | 1/2 | 1/2 | 1/2 | 1/20 |
| $s_5$ | 1/5 | 1/5 | 9/30 100 | 1/21 |
| $s_1$ | 4/30 | 5/30 100 | 1/5 111 | |
| $s_3$ | 3/30 1000 | 4/30 101 | | |
| $s_2$ | 2/30 1001 | | | |

**Figure 4.4   Fourth step of Huffman coding.**

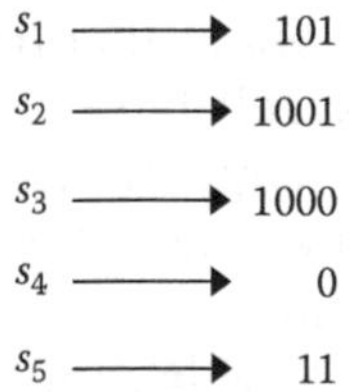

$$s_1 \longrightarrow 101$$
$$s_2 \longrightarrow 1001$$
$$s_3 \longrightarrow 1000$$
$$s_4 \longrightarrow 0$$
$$s_5 \longrightarrow 11$$

**Figure 4.5   Huffman codes.**

If we use block codes instead of Huffman codes, three bits are needed to represent five messages. Moreover, the entropy can be calculated for these probabilities.

$$H = 4/30 \log_2(30/4) + 3/30 \log_2(30/3) + 2/30 \log_2(30/2) + 1/5 \log_2(5) + 1/2 \log_2(2) = 1.35 \text{ bits}$$

If the probabilities were the power of two, the entropy should have the same value as the average code length. Huffman codes are the most efficient codes while the probabilities are the power of two.

## 4.2 Channel Coding

The theoretical basis for channel coding, which has come to be known as information theory, was stated by Claude Shannon in 1948. He showed that it is possible to achieve reliable communications over a noisy channel if the source's entropy is lower than the channel's capacity by defining the entropy of an information source and the capacity of a communications channel. However, Shannon noted that the channel capacity could be attainable, but he did not explicitly state how. At the same time, Hamming and Golay were developing practical error control schemes.

In the following subchapters, block codes, algebraic codes, cyclic codes, Reed Solomon codes, convolutional codes, recursive systematic convolutional codes, trellis coded modulation, and Viterbi algorithm will be briefly explained as in [1–12].

### 4.2.1 Block Codes

Block code is a type of channel coding that is used to combat noisy channel. The process of reducing the error probability or increasing the error performance is called channel coding. Block codes are one of the most accepted channel coding methods in literature. The memory size of the coding method and the signal constellation are very important. Furthermore, the noisy channel coding theorem, which was introduced by Claude Shannon, indicates that reliable communications can be provided for low signal-to-noise ratios if large random codes are used. However, it is not practical to use random codes, because the codes must have a structure for decoding algorithms.

Block code is a set of rules which the acceptable combinations of a certain number of message parts are assigned as code words. The transmitted bit stream must be uniquely decipherable in the receiver. As an example, it is assumed that 1, 01, 001, and 0001 are sent for four different messages. Prefix restriction is agreed on this example and thus, these messages are instantaneously decipherable. Even this is sufficient for a code to be uniquely decipherable, but not necessary. As another example, 1, 10, 100, and 1000 codes are assumed to be sent. Here, the prefix restriction is not agreed upon. In other words, the beginning bit or bits originate from other messages. In this case, it is not possible to decode the incoming message without considering the next bits. Therefore, these code words are not instantaneously decipherable [1–4].

Grouping symbols into blocks of $k$ and then adding $(n–k)$ redundant parity check symbols to produce an $n$ symbol code word is the general strategy of block coding. The code rate of the resulting code $R$ is $k/n$.

### 4.2.1.1 Code Distance

The number of different bits in two same-sized codes is called code length or Hamming distance. Code length is very important to correct errors while communicating over noisy channels. The probability of correcting the noisy code words received is as much as the code length. To make the code length larger, the length of the code word is increased. Therefore, the redundant bits are implemented behind the code words. The minimum code distance among all code words is denoted as $D_{min}$. On the receiver side, $D_{min}-1$ bit errors can be sensed and, $D_{min}/2-1$ bit errors can be corrected for an even number of errors and $D_{min}/2-1/2$ bit errors can be corrected for an odd number of errors.

### 4.2.1.2 Code Length

Code length is an expression that indicates the number of bits for code words. There is an inequality to correct the number of errors $E$ for the number of message words $M$ with the code length of $n$ bits.

$$M \leq \frac{2^n}{\displaystyle\sum_{i=0}^{E} \binom{n}{i}} \tag{4.1}$$

where

$$\binom{n}{i}$$

is a combination of $n$ for $i$. We can calculate the number of errors to be corrected for message words $M$ with $n$ bits or the minimum code length $n$ to correct the number of errors $E$ for message words $M$.

It is possible to recover more errors when the code length is increased. However, this causes a lowering of the code rate. Therefore, the percentage of the information bits in the code word and also the data transmission rate decrease.

### 4.2.1.3 Algebraic Codes

Algebraic codes are formed by appending the parity bits to the information bits. Parity bits are obtained using modulo 2 operations between the information bits. If there are $m$ bits in our information word, there should be a number of $2^m$ different information words. Here, the number of $n$ parity bits is appended to the message. Thus, the message consists of a number of $m + n$ bits. It is assumed that the information bits and parity bits are denoted as $a_i$ and $c_i$, respectively. The message bits are obtained by multiplying the information bits by the generator matrix **G**. The generator matrix includes $m \times m$ unit matrix **I** and $m \times n$ parity matrix **P** as below.

$$\mathbf{G} = \begin{bmatrix} \mathbf{I} & \mathbf{P} \end{bmatrix} \tag{4.2}$$

The output message is in the form of

$$\mathbf{M} = \begin{bmatrix} a_1 & a_2 & \cdots & a_m & c_1 & c_2 & \cdots & c_n \end{bmatrix} \tag{4.3}$$

At the receiver, the received message is multiplied by the transpose of check matrix **H**. The **H** matrix is in the form of

$$\mathbf{H} = \begin{bmatrix} \mathbf{P}^{\mathrm{T}} & \mathbf{I} \end{bmatrix} \tag{4.4}$$

The formula below is proved for noiseless received message words.

$$\mathbf{M}\mathbf{H}^{\mathrm{T}} = \mathbf{0} \tag{4.5}$$

As an example, **G** and its **H** matrix are given below for $m = 3$ and $n = 2$.

$$\mathbf{G} = \left[\begin{array}{ccc|cc} \overset{\mathbf{I}}{1} & 0 & 0 & \overset{\mathbf{P}}{1} & 0 \\ 0 & 1 & 0 & 0 & 1 \\ 0 & 0 & 1 & 1 & 1 \end{array}\right] \qquad \mathbf{H} = \left[\begin{array}{ccc|cc} \overset{\mathbf{P}^{\mathrm{T}}}{1} & 0 & 1 & \overset{\mathbf{I}}{1} & 0 \\ 0 & 1 & 1 & 0 & 1 \end{array}\right]$$

### 4.2.1.4 Cyclic Codes

Cyclic codes are generated by shifting a code word one bit at each coding step. If each $\mathbf{x} = (x_0, x_1, \ldots, x_{n-2}, x_{n-1})$ is an element of the block code set $C$ and also if $\mathbf{x}' = (x_{n-1}, x_0, x_1, \ldots, x_{n-2})$ is an element of the same set, $C$ is called *cyclic*. A generator matrix of cyclic codes is given below.

$$G = \begin{bmatrix} g_0 & g_1 & \cdots & g_{n-k} & 0 & \cdots & 0 \\ 0 & g_0 & g_1 & \cdots & g_{n-k} & \cdots & 0 \\ \vdots & & \ddots & \ddots & & \ddots & \vdots \\ 0 & \cdots & 0 & g_0 & g_1 & \cdots & g_{n-k} \end{bmatrix} \qquad (4.6)$$

Cyclic codes are generated by the polynomial of $g(D) = g_0 + g_1 D + g_2 D^2 + \ldots + g_{n-k} D^{n-k}$. Likewise, the message and the code can be indicated as a polynomial $m(D)$ and $x(D)$. Finally, the output code can be given as polynomial multiplication $x(D) = m(D)g(D)$. The polynomial multiplication is equivalent to the convolution computation. Thus, the output message is obtained as

$$x_i = \sum_{v=0}^{n-k} m_{i-v} g_v \qquad (4.7)$$

One of the most widely used polynomials for cyclic encoders is $g(D) = 1 + D + D^2$ and its encoder structure is shown in Figure 4.6.

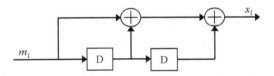

**Figure 4.6   Encoder structure for $g(D) = 1 + D + D^2$.**

### 4.2.1.5   Reed–Solomon Codes

Reed–Solomon (RS) codes are block-based error correcting codes used in wireless mobile communications, satellite communications, digital video broadcasting, and high-speed modems such as ADSL and xDSL. The Reed–Solomon encoder takes a block of $k$ data symbols of $s$ bits each and appends "redundant" $n-k$ parity symbols of $s$ bits each. The Reed–Solomon decoder can correct up to $t$ erroneous symbols, where $2t = n-k$.

As an example, a popular Reed–Solomon code RS(255,223) with 8-bit symbols is considered. The input of the encoder is 223 bytes and the encoder adds 32 parity bytes; thus, the output is 255 code word bytes. The maximum code word length $n$ for a symbol size $s$ can be calculated as

$n = 2s - 1$. Therefore, the maximum length of the code for the given example is 255 bytes.

Because the generator polynomial has $2t$ roots, there are only $2t$ parity checks. This is the lowest possible value for any $t$-error correcting code and is known as the Singleton bound. It should be constructed as an appropriate finite field and the roots should be also chosen to construct a generator for a Reed–Solomon code. The roots are generally in the form of $\alpha^i$, $\alpha^{i+1}$ to $\alpha^{i+2t-1}$, the generator polynomial of RS codes is given below:

$$g(X) = (X + \alpha^i)(X + \alpha^{i+1}) \cdots (X + \alpha^{i+2t-1}) \tag{4.8}$$

and the code word is encoded as

$$c(x) = g(x)i(x) \tag{4.9}$$

where $g(x)$ is the generator polynomial, $i(x)$ is the information block, $c(x)$ is the encode code word, and $\alpha$ is a primitive element of the field.

The encoding computation of Reed–Solomon code can be evaluated by a long division method or equivalently by a shift register network, which is shown in Figure 4.7. The shift register method is simpler and more applicable than the division method. In the binary case, the results of the calculations are always 0 or 1, but here it differs according to the appropriate value from the finite field.

For the example given above, there should be six registers that hold a symbol (8 bits). Here, finite field operations (addition and multiplication) on a complete symbol are considered.

Generally, the received code word is the addition of the transmitted code word and the noise as given below.

$$r(x) = c(x) + e(x) \tag{4.10}$$

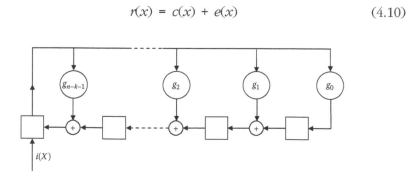

**Figure 4.7   Reed–Solomon encoder.**

In the decoding process, syndrome calculation is done first, which is a similar to parity calculation. There are $2t$ syndromes that depend only on errors and these can be calculated by substituting the $2t$ roots of the generator polynomial $g(x)$ into $r(x)$.

Then, the error locations of the symbols are found by solving $t$ simultaneous equations. There are two steps of this process. The first one is to find an error locator polynomial using Berlekamp–Massey or Euclid's algorithms. Euclid's algorithm is more widely used in practice. The second step is to find the roots of this polynomial using the Chien search algorithm. Another widely used fast algorithm for solving simultaneous equations with $t$ unknowns is the Forney algorithm.

### 4.2.1.6   Convolutional Encoders

Convolutional encoders are similar to cyclic encoders according to memory usage. However, cyclic codes are block codes and their code lengths are definite. Convolutional codes are produced by the input of continuous bit stream. Therefore, it is not meaningful to talk about the constant code length. Another difference is that although the cyclic encoders can be assumed as a linear system with one input and one output, convolutional encoders have multiple inputs and multiple outputs.

Convolutional encoders have a code rate as in block codes and given as $r = k/n$ but with a different meaning. $k$ is the number of input bit streams and $n$ is the number of output bit streams. The input message $\mathbf{m}$ consists of the number of $k$ bit streams or can be divided into $k$ bit streams. The number of $k$ bit input streams is passed through the encoder and the number of $n$ output bit streams, $\mathbf{x}^{(q)}$ $q \in (0,\ldots n - 1)$, is transmitted.

Convolutional codes are produced by the number of $n \times k$ generator polynomial $g^{(p,q)}(D), p \in (0,\ldots k-1)$ and $q \in (0,\ldots n-1)$. There are a number of $k$ shift registers for each input bit stream. The number of memory in the shift register for the $p$th input is denoted as $M_p$. The total memory size can be given as

$$M_c = \sum_{p=0}^{k-1} M_p \tag{4.11}$$

Constraint length can be defined as the maximum number of bits in the output bit stream that are affected by any input bit and can be formulized as

$$K_c = 1 + \max_p M_p \tag{4.12}$$

where maxM is the maximum degree of the concerned polynomial. The coding process continues by convolving each input bit stream with its related polynomial. This convolution is given in the following equation.

$$w_i^{(p,q)} = \sum_{v=0}^{M_p} m_{i-v}^{(p)} g_v^{(p,q)} \tag{4.13}$$

The output of the convolutional encoder, $\mathbf{x}^{(q)}$, is

$$\mathbf{x}^{(q)} = \sum_{p=0}^{k-1} \mathbf{w}^{(p,q)} \tag{4.14}$$

If we assume that there is only one input bit stream, in other words $k = 1$, the code rate would be $r = 1/n$. Therefore, $p = 0$ and is omitted. In Figure 4.8, a conventional convolutional encoder is shown for $g^{(0)} = 1 + D + D^2$ and $g^{(1)} = 1 + D^2$ with the code rate $r = 1/2$. In general, coefficients of polynomials are given in octal form. The coefficients of the encoder shown in Figure 4.8 are (7,5).

A convolutional encoder is a finite state Markov process that can be represented by a state diagram and a trellis structure. The state of the encoder is determined by the contents of the memories. There are a number of $2^{M_c}$ states for the number of $M_c$ memories. The states of the memories change according to the value of an input bit. Then, the encoder produces an $n$ bit output.

The state diagram is more understandable for the function of convolutional encoder. The states are denoted as $S$ and there are number of $2^{M_c}$ states in the state diagram. All the states are connected over the branches and the expression of $m/\mathbf{x}$ is on those branches, where $\mathbf{x}$ is the output for the input of $m$. The state diagram is illustrated in Figure 4.9 for the convolutional encoder in Figure 4.8.

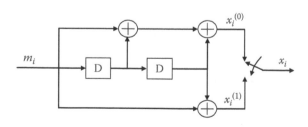

**Figure 4.8  Convolutional encoder for $r = 1/2$ and $K_c = 3$.**

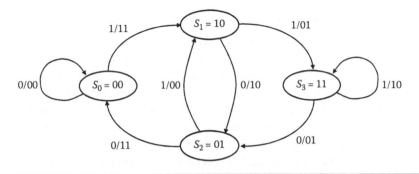

**Figure 4.9 State diagram of the convolutional encoder shown in Figure 4.8.**

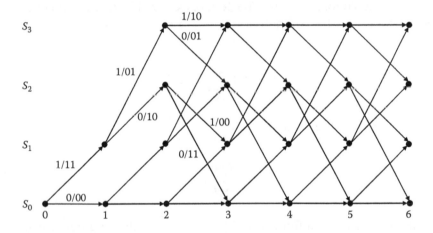

**Figure 4.10 Trellis structure of the convolutional encoder shown in Figure 4.8.**

The state diagram can be shown with a trellis structure for each coding step. In Figure 4.10, the trellis structure is given for six coding steps. In convolutional encoders, each code word is represented by a branch in the trellis. This branch is called a state sequence and is denoted as $\mathbf{s} = (s_0 \ldots , s_{2Mc})$. At the beginning of the coding, the state of the memories is zero; thus, the first state of the trellis is zero. There is a one-to-one correspondence between the state sequence $\mathbf{s}$, the code word $\mathbf{x}$ of that state sequence, and the message $\mathbf{m}$.

The minimum distance of the convolutional code is called the minimum free distance and is denoted as $d_{free}$. This can be explained as $d_{free}$ is the smallest Hamming weight of all the code words that are related to the branches $\mathbf{s}$, which come from the state zero at coding step 0. The free distance is the most important performance criterion for convolutional codes.

$$d_{free} = \min\{w(\mathbf{x}) \text{ for } x_0 = 1\} \qquad (4.15)$$

### 4.2.1.7 Recursive Systematic Convolutional Codes

Convolutional codes are made systematic without reducing the Hamming distance as are cyclic codes. In Figure 4.11, a rate of 1/2 convolutional code is given. First, the remainder $r(D)$ of the polynomial division $m(D)/g^{(0)}(D)$ is calculated. Then, the parity output polynomial is calculated by the polynomial multiplication $x^{(1)}(D) = r(D)g^{(1)}(D)$. The systematic output is the same as the input $x^{(0)} = m(D)$. The remainder $r(D)$ is computed using a recursive shift register network, where $g^{(0)}(D)$ is called the feedback polynomial and $g^{(1)}(D)$ is called the feedforward polynomial. This type of encoder is called a recursive systematic convolutional (RSC) encoder and codes generated in this encoder are called recursive systematic convolutional codes. The feedback variable or the remainder of RSC encoder can be computed as

$$r_i = m_i + \sum_{v=1}^{M_C} r_{i-v} g_v^{(0)} \tag{4.16}$$

and then finding the parity output

$$x_i^{(1)} = \sum_{v=1}^{M_C} r_{i-v} g_v^{(1)} \tag{4.17}$$

RSC encoders are a finite state Markov process that can be represented by a state diagram and a trellis structure. The code rate of the RSC encoder shown in Figure 4.11 is 1/2 and $K_c = 3$. The state and the trellis diagrams of this RSC encoder are shown in Figures 4.12 and 4.13, respectively. The state and trellis diagrams for the RSC code obtained from Figure 4.11 and the related nonsystematic convolutional code obtained from Figure 4.8 are

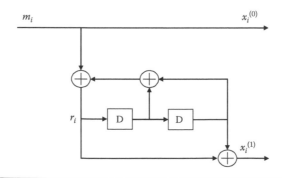

**Figure 4.11** RSC encoder for rate 1/2 and $K_c = 3$.

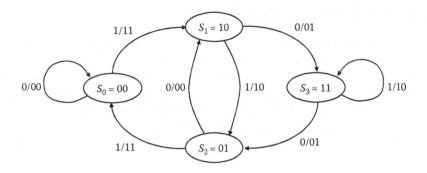

**Figure 4.12 State diagram of RSC encoder for rate 1/2 and $K_c = 3$.**

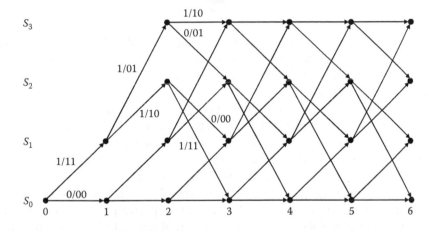

**Figure 4.13 Trellis diagram of RSC encoder for rate 1/2 and $K_c = 3$.**

almost identical. The only difference between these two state diagrams is that the input bits labeling the branches leaving states $S_1$ and $S_2$ are different. The minimum free distance remains the same, because the structure of the trellis and the output bits is the same.

In general, the state of the encoder is made zero at the end of each frame. Hence, the state of each consecutive frame can start from state 0. In convolutional encoders, the trellis can be forced back to zero by padding the message with the number of $M_c$ zero bits. However, it is not the same for RSC codes because of its recursive property. Therefore, the message input $m_i$ should be chosen such that $r_i = 0$ for the last $M_c$ coding step. Thus, this input message is called dummy bits and can be computed as

$$m_i = \sum_{v=1}^{M_c} r_{i-v} g_v^{(0)} \tag{4.18}$$

## 4.3 Trellis-Coded Modulation

Trellis-coded modulation (TCM) is the joint form of coding and modulation introduced by Ungerboeck in 1982 [8]. In the TCM system design, complex signal values are placed instead of their binary codes on the trellis structure. If we assume that $a_i$ and $b_i$ are the channel signals, the squared Euclidean distance is denoted as $d^2(a_i, b_i)$. The minimum squared distance $d^2_{free}$ is the smallest squared distance between the channel signal streams that leave from one state and join after one or more transitions. $d^2_{free}$ can be calculated as

$$d^2_{free} = \min \sum_{a_n \ a'_n} d^2(a_n, a'_n) \qquad (4.19)$$

In Figure 4.14, the input symbol is composed of $n$ bits and $n + 1$ bits exit from the convolutional encoder with the rate $n/(n + 1)$. Then, the output bits are mapped to $2^{n+1}$ channel signals using the rule named mapping by partitioning.

According to the rule of mapping by partitioning, the set of excessive channel signals is partitioned into subsets while the free Euclidian distance increases in certain symmetry. Then, these subsets are designated to the state transitions of the chosen trellis structure as the free Euclidian distance between the channel signal streams is maximized.

In the receiver, generally the Viterbi algorithm is utilized for decoding. If a soft decision decoder is used, better error performance should occur. The partitioning of 8-PSK signals is shown in Figure 4.15 [8]. Here, $\Delta_i$ is the squared distance between two signals that are in the same subsets. In the rule of mapping by partitioning, the $\Delta_0 < \Delta_1 < \Delta_2$ condition is stipulated.

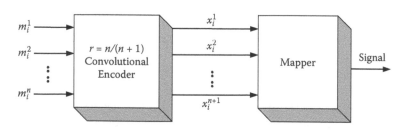

**Figure 4.14   Trellis-coded modulation (TCM) block diagram.**

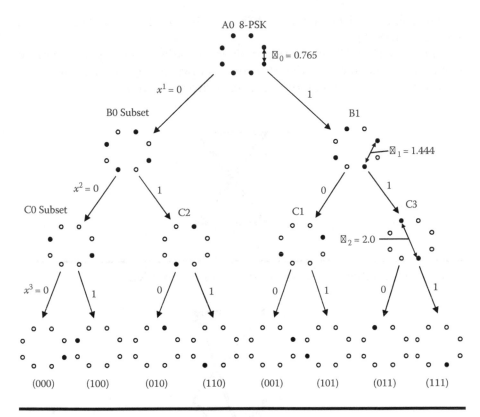

**Figure 4.15    Partitioning of an 8-PSK signal set.**

Two, four, and eight state 8-PSK signals are shown in Figures 4.16, 4.17, and 4.18 [8].

Figures 4.16, 4.17, and 4.18 are given according to the partitioning in Figure 4.15, and the encoder structures shown in Figures 4.19 and 4.20 have two and three memory units, respectively as in [8].

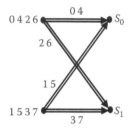

**Figure 4.16** **An 8-PSK signal set using a two-state encoder.**

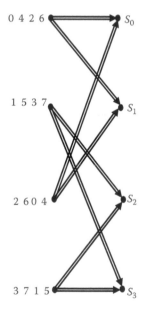

**Figure 4.17** **An 8-PSK signal set using a four-state encoder.**

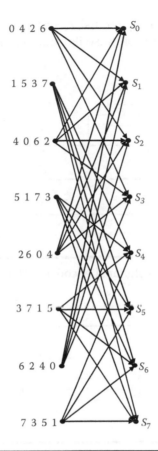

**Figure 4.18  An 8-PSK signal set using an eight-state encoder.**

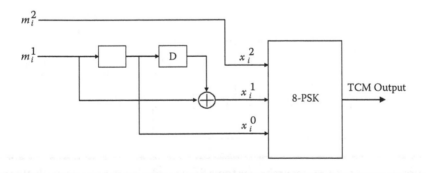

**Figure 4.19  An 8-PSK convolutional encoder with two memories.**

In Figures 4.21 and 4.22, recursive systematic convolutional encoder structures are given instead of those illustrated in Figures 4.19 and 4.20 [8].

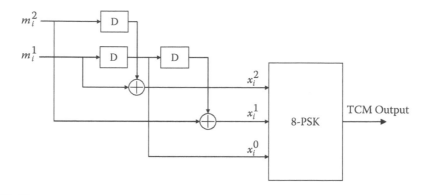

**Figure 4.20** **An 8-PSK convolutional encoder with three memories.**

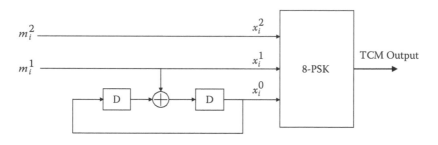

**Figure 4.21** **An 8-PSK recursive systematic convolutional encoder with two memories.**

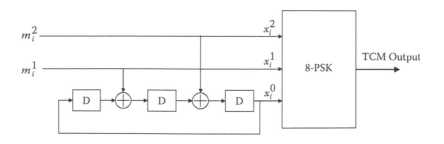

**Figure 4.22** **An 8-PSK recursive systematic convolutional encoder with three memories.**

## 4.4 Viterbi Algorithm

The Viterbi algorithm is the most common and widespread decoding algorithm for convolutional encoders. It is used in most systems as in [2–5,11,12]. In this algorithm, the continuous path on the trellis diagram is searched during a certain data frame; the path is chosen that minimizes the Gaussian density function. Thus, the states of the trellis structure are obtained for the related frame. After finding the states, the message bits can be estimated without difficulty.

First, the received noisy signal is compared to all signals according to the Gaussian density function as in Equation 4.20. Then, the metric values for each probable signal are computed.

$$P(s_n) = \frac{1}{\sqrt{2\pi}\sigma} \exp\left(\frac{(-s_n - s')^2}{2\sigma^2}\right) \tag{4.20}$$

where $s_n$ is the probable noiseless signal, $\sigma^2$ is the variance of the noise, and $s'$ is the received noisy signal. Subscript $n$ varies between 0 and 7 for 8-PSK signaling. For each received noisy signal, the number of $n$ metric values is computed using Equation (4.20). After this operation, for each state, metric values are mapped to the state transition, causing this state to occur and probability values of previous states, related to metric values, are added and the minimum is chosen.

At the end of the frame, the path on the trellis, which minimizes the Gaussian function, is found. The input messages can be determined from these states.

To simplify the metric computation most of the time only $s_n - s'^2$ is used. At the same time, if the fading parameters are known for each received symbol in wireless fading channels, to compute more precise metric values, the received signal is divided into the fading parameters, $s_n - s'\rho^2$, or the probable noiseless symbol is multiplied by the fading parameter, $\rho.s_n - s'^2$.

## References

[1] Oppenheim, A. V., and R. W. Schafer. *Digital Signal Processing*. Upper Saddle River, NJ: Prentice Hall, 1975.

[2] U•an, O. N., and A. Muhittin Albora. *New Approaches in Signal and Image Processing* (in Turkish). Istanbul: Istanbul University, 2003.

[3] U•an, O. N., and O. Osman. *Computer Networks and Communication Techniques* (in Turkish). Istanbul: Istanbul University, 2003.

[4] U•an, O. N. Error Performance of Quadrature Partial Response Trellis Coded Modulation (QPR-TCM). Ph.D. Thesis, Istanbul Technical University, 1994.

[5] Osman, O. Performance of Turbo Coding Techniques. Ph.D. Thesis, Istanbul University, 2004.

[6] Odabasioglu, N. Error Performance of Continuous Phase Source/Channel Coding in Fading Channels. MSc Thesis, Istanbul University, 2002.

[7] Valenti, M. C. Iterative Detection and Decoding for Wireless Communications. Ph.D. Thesis, Virginia Polytechnic Institute and State University, Blacksburg, Virginia, 1999.

[8] Ungerboeck, G. Channel Coding with Multilevel/Phase Signals. *IEEE Trans. on Information Theory* IT-28(1):55–67, Jan., 1982.

[9] Proakis, J. G. *Digital Communications,* 4th Edition. New York: McGraw-Hill, 2001.

[10] Roden, M. S. *Digital and Data Communication Systems.* Upper Saddle River, NJ: Prentice Hall, 1982.

[11] Viterbi, A. J. Error Bounds for Convolutional Codes and an Asymptotically Optimum Decoding Algorithm. *IEEE Transactions on Information Theory* 13(2):260–269, April 1967.

[12] Forney, G. D. The Viterbi Algorithm. *Proceedings of the IEEE* 61(3):268–278, March 1973.

# Chapter 5

# Channel Models

In mobile channels, transmitted signals reach the receiver by moving over different paths. There should be either a line of sight or different ground shapes such as buildings, hills, or planet covers between the transmitter and receiver. Because of the ground shapes, reflection, refraction, or scattering could cause electromagnetic waves to be emitted during mobile communications. The transmitted signals are delayed reaching the receiver because of these obstacles. Mobile radio channels change randomly and, therefore, are difficult to investigate, unlike stationary channels. Transmitted symbols reach the receiver over various channels with different time intervals. The case in which signals are transmitted to the receiver over various paths is called a multipath channel.

## 5.1 Mobile Communication Channels

The most frequently used and the simplest channel model is the additive white Gaussian noise (AWGN). In this model Gaussian noise is added to the received signal. The AWGN channel is very important in communications theory and practice particularly for deep space and satellite communications. However, in most of the channels, different fading effects are encountered. In such cases, AWGN channels become insufficient and more suitable channel models are needed. In a free-space model, it is assumed that there is no object between the transmitter and receiver, the atmosphere is completely uniform, and the reflection coefficient is nearly zero for the spread signals. Nevertheless, the free-space model is not capable of modeling the channels where communication is close to the ground.

In mobile communications, the transmitted signal can be shown as

$$\mathbf{x} = (x_1, x_2, \ldots, x_N) \qquad (5.1)$$

where $x_k = \sqrt{2E_s}\,e^{j\phi_k}$, $\phi_k = 0,\ 2\pi/M,\ \ldots,\ 2(M-1)\pi/M$ for m-ary phase shift keying (MPSK) signals. Corresponding to $\mathbf{x}$, a noisy discrete-time sequence $\mathbf{y} = (y_1, y_2, \ldots, y_N)$ appears at the output of the channel, where $y_k$, $k=1$, 2, ..., $N$, is given by

$$\mathbf{y} = \mathbf{H} * \mathbf{x} + \mathbf{n} \qquad (5.2)$$

where $\mathbf{x}$ and $\mathbf{n}$ are the modulated signal sequence and additive white Gaussian noise, $\mathbf{H}$ is a vector that includes the channel coefficients, and * is the convolutional computation.

## 5.2  Fading Channels

Fading channels are the most common channel models in mobile radio communications. In mobile communications channels, because the transmitted signals are generally sent to the receiver over two or more paths, interference occurs, seriously affecting system performance. The received multipath signals have different phases and amplitudes because of the delay and reflections. The variation of the amplitude and the phase shift of the received signal can change very quickly within a very extensive range. If there are many paths that form reflections, and also if there is no line of sight, an envelope of the received signal could be modeled as a Rayleigh probability distribution function (pdf) as

$$P(\rho) = 2\rho e^{-\rho^2}, \quad \rho \geq 0 \qquad (5.3)$$

If there is a distinct line of sight in addition to the reflections, the received signal is modeled as a Rician pdf where the amplitude of the signal received from the line of sight becomes lower, the Rician pdf turns into a Rayleigh pdf.

$$P(\rho) = 2\rho(1+K)\ e^{(-\rho^2(1+K)-K)} I_0\left[2\rho\sqrt{K(1+K)}\right], \quad \rho \geq 0 \qquad (5.4)$$

where $I_0$ is the zero-order modified Bessel function of the first kind, and $K$ is the Rician parameter, defined as the ratio of the direct power of the specular components to the power of the diffuse component. Rayleigh fading corresponds to the limiting case of the Rician channel where $K = 0$.

Each path, including reflections, can cause a different power loss, delay, and phase shift. The increment on the signal amplitude is observed if reflected signals are received with the same phase. On the contrary, if the signals are received with different phases, the decrement on the amplitude is observed. Consequently, the received signal is the resultant power of the signals, which come from different channels [1].

Change in communication channels, such as rain or snow, causes fading. Doppler spread is another factor that affects a correct determination of the received signal. It causes a frequency shift on the carrier and thus, an increase in the spectrum.

The channel with fading effect and imperfect phase can be modeled as

$$b = \rho \, e^{j\theta} \tag{5.5}$$

where $\rho$ is fading amplitude and the term $e^{j\theta}$ is a unit vector where $\theta$ represents the phase noise as mentioned in [2–5], which is assumed to have a Tikhonov pdf given by

$$p(\theta) = \frac{e^{\eta\cos(\theta)}}{2\pi I_0(\eta)} \qquad |\theta| \le \pi \tag{5.6}$$

where $I_0$ is the modified Bessel function of the first kind order zero and $\eta$ is the effective signal-to-noise ratio (SNR) of the phase noise. The received signal for nonfrequency-selective fading channels can be written as

$$y_k = \rho_k x_k e^{j\theta_k} + n_k \tag{5.7}$$

## 5.2.1 Large-Scale and Small-Scale Fading

Fading can be classified into two categories: large- and small-scale fading. Large-scale fading is defined as an average weakening of the power or a path loss. This type of fading occurs when there are considerable ground shapes between the receiver and transmitter, such as a hill, forest, or mass of buildings. In the case of small-scale fading, alterations can occur on the amplitude or on the phase of the transmitted symbols. In mobile radio applications, the fact that movement between transmitter and receiver results in some difference in the signal paths, means the channel becomes a time varying channel. Spreading speed also influences the fading effect on the signal. This effect can be minimized by making some adjustment to the bit transfer rate according to the variation of the channel in time domain.

## 5.2.2 Mathematical Model of Multipath Fading Channels

Various parameters, such as delay spread, bandwidth, etc. can be used to explain multipath fading channels. Any multipath signal can be written as

$$x(t) = \text{Re}\left\{ x'(t)e^{j2\pi f_c t} \right\}$$
(5.8)

where $f_c$ is the carrier frequency and $x'(t)$ is the complex envelope of the transmitted signal. The signal detected at the receiver block is given by Equation 5.9. It is assumed the channel is composed of $N$ paths as

$$y(t) = \text{Re}\left\{ \sum_{n=1}^{N} \rho_n x'(t-\tau_n)e^{j2\pi\left[(f_c+f_{D,n})(t-\tau_n)\right]} \right\}$$
(5.9)

where $\tau_n$ is time delay, $f_{D,n}$ is the Doppler shift and $\rho_n$ is fading parameter. Then the received bandpass signal complex envelope is

$$y'(t) = \sum_{n=1}^{N} \rho_n x'(t-\tau_n)e^{-j\theta_n(t)}$$
(5.10)

where $\theta_n$ is the phase of $n$th propagation path

$$\theta_n(t) = 2\pi\left[(f_c + f_{D,n})\tau_n - f_{D,n}t\right]$$
(5.11)

### 5.2.2.1 Compatibility

Bandwidth compatibility, $B_c$, defines the degree of passing signals through a channel with the same fade and phase distortion. The mathematical expression of bandwidth compatibility is given as

$$B_c = \frac{1}{2\pi\sigma_\tau}$$
(5.12)

where $\sigma_\tau$ is the standard deviation of the additional delay.

## 5.2.2.2  Doppler Spread

In the case of relative movement between transmitting and receiving antennas, a constant frequency shift occurs called $f_d$, Doppler frequency. $B_d$ is also a related measurement of the spectral enlargement because of this movement or the movements of the objects; this bandwidth is equal to the maximum Doppler spread $B_d = f_d$. If the bandwidth of the baseband signal is large enough, according to the Doppler shift, the Doppler shift may be discarded. Doppler frequency of any path can be written as

$$f_d = f_c \frac{v}{c} \qquad (5.13)$$

where $f_c$ is the carrier frequency, $v$ is the relative speed between transmitter and receiver, and $c$ is the light speed. Doppler spread is the maximum frequency shift that can occur in different paths. Total bandwidth is the summation of baseband signal bandwidth and Doppler shift. The Doppler period $T_d$ is the inverse of the Doppler frequency, $T_d = 1/f_d$.

## 5.2.2.3  Delay Scattering

Scattering is one of the main reasons for delay because of signal multipath traveling. The time interval of multipath delay for $i$th path is defined as $\tau_i$. For any path, this delay differs, but for the overall system, an average delay occurs denoted as $\bar{\tau}$. If multipath channel scattering is closer to the ideal, then $\bar{\tau}$ converges to zero. As in all stochastic models, the square root of variance, $\sigma_\tau$, called the standard deviation, is another important parameter to be considered. The standard deviation is microseconds for space radio communications, and nanoseconds for indoor communication cases.

## 5.2.2.4  Compatibility Duration

Compatibility duration $T_c$ is the time interval in which the interference occurs. $T_c$ can be written approximately as a function of Doppler spread as

$$T_c \approx \frac{9}{16\pi f_d} \qquad (5.14)$$

## 5.2.3  *Fading Channel Types*

Fading channels are classified according to the bandwidth compatibility $B_c$ and signal bandwidth $B_s$ relation. If $B_c$ is greater than $B_s$, the channel is

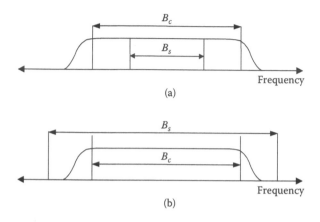

**Figure 5.1 Fading-channel characteristics: (a) flat-fading channel, (b) frequency-selective channel.**

called flat, otherwise the channel becomes frequency selective as shown in Figure 5.1.

### 5.2.3.1 Flat Fading

If the bandwidth compatibility $B_c$ of a wireless channel is greater than the signal bandwidth $B_s$, the fading occurring on the signal is called a flat-fading or a nonfrequency selective channel. In flat fading, the gain and linear phase are constant, thus spectral properties of the transmitted signal are preserved on the receiver side. In mathematical representation, flat-fading channels have a characteristic of

$$B_c \gg B_s \tag{5.15}$$

Because frequency response of a fading channel is flat, the impulse response of the flat-fading channel is a delta function without any delay. Moreover, the amplitude of the delta function changes in time.

### 5.2.3.2 Frequency Selective Channels

If the bandwidth of the channel $B_c$-constant gain and linear phase property is less than the bandwidth of the signal $B_s$, the frequency-selective channel is in question. The received signals are similar to faded transmitted signals with different time delays. Because of scattering, the transmitted signal and delayed symbols interfere with each other in the time domain. When this is investigated in the frequency domain, it reveals that amplitudes of some frequencies are stronger in the spectra. The models of these channels are quite complex when compared to the flat-fading channels. Assume that each multipath signal is modeled as if single path with linear filter characteristics. For frequency-selective channels, the following equation indicates the frequency-selective channel characteristic.

$$B_c < B_s \tag{5.16}$$

According to Equation (5.16), in frequency-selective channels, bandwidth compatibility is greater than the signal bandwidth, or equivalently, the square root of the square mean of delay spreads is greater than the symbol period. In practice these criteria can be modified according to the modulation type. In conclusion, if the $\sigma_\tau$ is greater than $T_s$, the channel is called a frequency-selective channel.

### 5.2.3.3  Slow-Fading Channel

In slow-fading channels, the parameters remain constant during one symbol transmission time. The channel compatibility duration is longer than the symbol period. In other words, the Doppler spread is less than the signal bandwidth. These can be shown as follows:

$$T_s << T_c \tag{5.17}$$

$$B_s >> B_c \tag{5.18}$$

### 5.2.3.4  Fast Fading

If the impulse response of the fading channel alters in one symbol duration, the channel is called fast fading; otherwise, the channel is slow fading. Instant variations occur because of the movements or Doppler spread. For fast-fading channels the following equation shows the channel characteristic from a period and bandwidth point of view.

$$T_s > T_c \tag{5.19}$$

or, equivalently,

$$B_s < B_c \tag{5.20}$$

The classification of a fading channel is illustrated in Figure 5.2. This classification helps to model the channels.

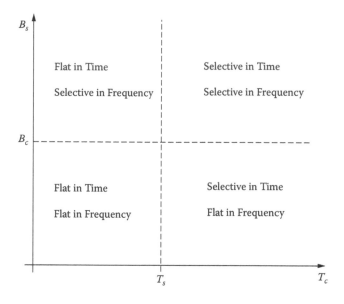

**Figure 5.2  Classification of fading channels.**

If a slow data rate and a high relative velocity between transmitter and receiver are taken into account, fast fading affects the signal. Naturally, if a high data rate and slow relative velocity are considered, fading will be slow. In the case of zero mobility, fading will be slow and be independent from the data rate.

There are some crucial points regarding modeling the channel. To know whether the channel is fast fading or slow fading doesn't provide any clue about whether the channel is frequency selective or flat. The impulse response of a fast flat-fading channel changes faster than symbol duration. For a fast frequency-selective fading channel, phase amplitude or delay in some frequencies changes faster than symbol duration. The other two channel types are slow flat-fading and slow frequency-selective channels. If we think that slow fading occurs in the case of fast data rates, it is more likely to be encountered in the real world.

## 5.2.4  Static Model of Fading Channels

The delay scattering of a flat-fading channel is lower than symbol duration. Therefore,

$$x(t - \tau_k(t)) \approx x(t) \tag{5.21}$$

The received noiseless signal is

$$y(t) = x(t) \sum_{n=1}^{N} \rho_n(t) e^{j\theta_n(t)} \qquad (5.22)$$

Thus, the static model of the multipath channel is given in (5.23) [17].

$$\alpha(t) = \sum_{n=1}^{N} \rho_n(t) e^{j\theta_n(t)} = a(t) + jb(t) \qquad (5.23)$$

In this case, $a(t)$ and $b(t)$ are given in (5.24) and (5.25) [17].

$$a(t) = \sum_{n=1}^{N} \rho_n(t) \cos(\theta_n(t)) \qquad (5.24)$$

$$b(t) = \sum_{n=1}^{N} \rho_n(t) \sin(\theta_n(t)) \qquad (5.25)$$

In most of the studies; $a(t)$ and $b(t)$ are assumed to have a Gaussian distribution.

## 5.3 Frequency-Selective Multipath Fading

If the time domain duality of signal bandwidth is much greater than the time spread of the propagation path delay, then channel models can be suitable for narrow band transmission. This means that the duration of a modulated symbol is much greater than the time spread of the propagation path delays. As such, all the frequencies in the transmitted signal are affected from the same random attenuation and phase shift due to multipath fading. This kind of channel is called flat fading and causes very little or no distortion to the received signal.

If the spread of the propagation path delays is larger than the frequency response of the channel, there may be a rapid change on the frequency scale comparable to the signal bandwith. In this case, fading is said to be frequency selective; the channel introduces amplitude and phase distortion to the transmitted signal. To flatten and linearize the signals, which are

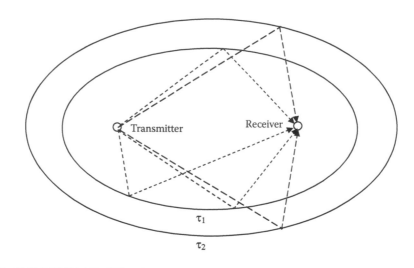

**Figure 5.3   Path geometry for a multipath fading channel.**

affected by the frequency selective channel, requires adaptive equalizers. The path geometry for a multipath-fading channel is shown in Figure 5.3 as in [17]. In this figure, single reflections are shown.

The impulse response of a complex low-pass channel depends on the complex input $x(t)$ and output $y(t)$ [17]. The output is the convolution of the input and channel function.

$$y(t) = \int_0^\infty x(t - \tau)\, b(t, \tau)\, d\tau \qquad (5.26)$$

The impulse response of a low-pass channel is expressed with input delay-spread function $b(\tau, t)$. This indicates the response of the channel at time $t$ for an impulse applied at time duration of $t - \tau$. Therefore, for this physical channel, $b(\tau, t) = 0$ and the lower base of integration in Equation (5.26) will be zero. The discrete time for the previous equation can be given as

$$y(t) = \sum_{m=0}^{n} x(t - m\Delta\tau)\, b(t, m\Delta\tau)\, \Delta\tau \qquad (5.27)$$

Multipath-fading channels can be modeled as a combination of various time variant linear filters [6] as shown in Figure 5.4.

In the frequency domain of channel-related function $H(f,v)$ called the Doppler-spread function, it depends on $X(f)$ and $Y(f)$ as shown in Figure 5.5.

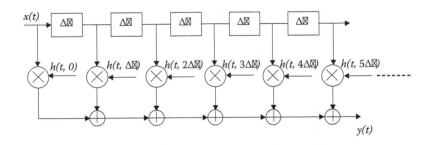

**Figure 5.4   Tap delay multipath fading channel model in discrete time.**

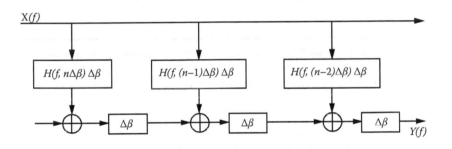

**Figure 5.5   Multipath fading-channel model in frequency domain.**

The frequency spectrum of the channel output is

$$Y(f) = \int_{-\infty}^{\infty} X(f-\beta)H(f-\beta,\beta)d\beta \qquad (5.28)$$

where $\beta$ is the Doppler shift. In discrete representation, Equation (5.28) turns into additions in discrete time as [17].

$$Y(f) = \sum_{m=0}^{n} X(f-m\Delta\beta)H(f-m\Delta\beta,m\Delta\beta)\Delta\beta \qquad (5.29)$$

Fourier transformations of these four functions are shown in Figure 5.6.

The third function, $T(f,t)$, depends on the integral of input signal spectrum and waveform of the output signal [17] and this function is called the time variant transfer function by Zadeh [8].

$$y(t) = \int_{-\infty}^{\infty} X(f)T(f,t)e^{j2\pi ft}\, df \qquad (5.30)$$

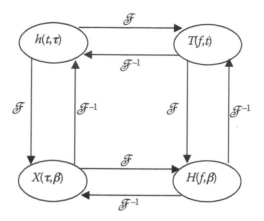

**Figure 5.6  Fourier transformations between the transmission functions.**

Delay Doppler-spread function $X(\tau,v)$ is the last transmission function and its effect is given as

$$y(t) = \int_{-\infty}^{\infty} \int_{-\infty}^{\infty} x(t-\tau) X(\tau,\beta) e^{-j2\pi f\tau} d\beta\, d\tau \qquad (5.31)$$

$X(\tau,\beta)$ explains the amplitude of channel spread with time delay $\tau$ and Doppler shift $\beta$ [6,17].

## 5.3.1  Correlation Functions of Statistical Channels

The impulse response of the channel $h(t,\tau)$ can be defined as a complex Gaussian random process $h(t,\tau) = h_I(t,\tau) + jh_Q(t,\tau)$. The quadratic components are inphase and quadrature phase. Both inphase $h_I(t,\tau)$ and quadrature phase $h_Q(t,\tau)$ are modeled as a Gaussian random process. To characterize the channel correctly, the combined probability density function (pdf) of the transmission functions must be known. However, it is not easy to obtain pdf information; therefore, the best approach is to find the statistical correlations for each transmission function. Clearly, statistical properties are best defined by means and autocorrelations. Assume the mean value to be zero, thus only the autocorrelation functions are used. In Figure 5.7, autocorrelation functions of four transmission functions, given in Equations (5.32)–(5.35), are defined as a function of the average of various filtering [1,7,17].

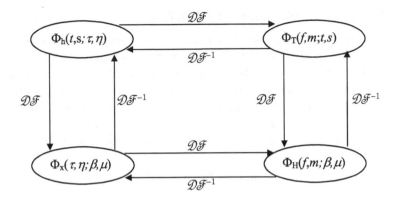

**Figure 5.7  Fourier transformations between channel autocorrelation functions.**

$$E\left[ b(t,\tau)b^*(s,\eta) \right] = 2 \ _b(t,x;\tau,\eta) \qquad (5.32)$$

$$E\left[ T(f,t)T^*(m,x) \right] = 2 \ _T(f,m;t,x) \qquad (5.33)$$

$$E\left[ H(f,\beta)H^*(m,\mu) \right] = 2 \ _H(f,m;\beta,\mu) \qquad (5.34)$$

$$E\left[ X(\tau,\beta)X^*(\eta,\mu) \right] = 2 \ _X(\tau,\eta;\beta,\mu) \qquad (5.35)$$

To calculate these autocorrelation functions, utilize the two-dimensional Fourier transformation shown in Equations (5.36) and (5.37) [17].

$$_X(\tau,\eta;\beta,\mu) = \int_{-\infty}^{\infty}\int_{-\infty}^{\infty} \ _b(t,s;\tau,\eta)\, e^{j2\pi(\beta t-\mu s)}\, dt\, ds \qquad (5.36)$$

$$_b(t,s;\tau,\eta) = \int_{-\infty}^{\infty}\int_{-\infty}^{\infty} \ _X(\tau,\eta;\beta,\mu)\, e^{-j2\pi(\beta t-\mu s)}\, d\beta\, d\mu \qquad (5.37)$$

## 5.3.2  *Classification of Statistical Channels*

There are three main classes of statistical channels. These are wide-sense stationary, uncorrelated scattering, and wide-sense stationary uncorrelated scattering.

### 5.3.2.1 Wide-Sense Stationary Channels

In wide-sense stationary (WSS) channels, fading statistics such as means, variance, etc. remain constant for short periods of time. In other words, a correlation function depends on the time period $\Delta t = s - t$ not the instant time parameters of $s$ and $t$. These channels are suitable for uncorrelated Doppler spreads. The most important property of this channel is that the Doppler shifts of the signals, which are affected by fading and phase shift, are uncorrelated. The correlation functions are as follows [17]:

$$_T(f,m;t,t+\Delta t) = \ _T(f,m;\Delta t) \tag{5.38}$$

$$_b(t,t+\Delta t;\tau,\eta) = \ _b(\Delta t;\tau,\eta) \tag{5.39}$$

$$_H(f,m;\beta,\mu) = \delta(v-\beta)\Psi_H(f,m;v) \tag{5.40}$$

$$_x(\tau,\eta;\beta,\mu) = \delta(\beta-\mu)\Psi_x(\tau,\eta;\beta) \tag{5.41}$$

Here, related Fourier transformation pairs are as follows [17]:

$$\Psi_H(f,m;\beta) = \int_{-\infty}^{\infty} {}_T(f,m;\Delta t)e^{-j2\pi\beta\Delta t}d\Delta t \tag{5.42}$$

and

$$\Psi_x(\tau,\eta;\beta) = \int_{-\infty}^{\infty} {}_b(\Delta t;\tau,\eta)e^{-j2\pi\beta\Delta t}d\Delta t \tag{5.43}$$

### 5.3.2.2 Uncorrelated Scattering

Uncorrelated scattering (US) channels are modeled by uncorrelated fading and phase shifts of different delay paths. If the frequency change of the correlation function of this channel is $\Delta f = m - f$, then the US channel becomes WSS [6]. Channel correlation functions include time delays that are singular as in Equations (5.44) and (5.47). Channel correlation functions of US channels can be given as [17]

$$_b(t,s;\tau,\eta) = \Psi_b(t,s;\tau)\delta(\eta-\tau) \tag{5.44}$$

$$_T(f,f+\Delta f;t,s) = \ _T(\Delta f;t,s) \tag{5.45}$$

$$_H(f,f+\Delta f;\beta,\mu) = \ _H(\Delta f;\beta,\mu) \tag{5.46}$$

$$_s(\tau,\eta;\beta,\mu) = \Psi_x(\tau;\beta,\mu)\delta(\eta-\tau) \tag{5.47}$$

Here, Fourier transformation pairs are as follows [17]:

$$\Psi_b(t,s;\tau) = \int_{-\infty}^{\infty} {}_T(\Delta f;t,s)e^{j2\pi\Delta f\tau}d\Delta f \tag{5.48}$$

and

$$\Psi_S(\tau;\beta,\eta) = \int_{-\infty}^{\infty} {}_H(\Delta f;\beta,\mu)\, e^{j2\pi\Delta f\tau}d\Delta f \tag{5.49}$$

### 5.3.2.3 Wide-Sense Stationary Uncorrelated Scattering

Wide-Sense Stationary Uncorrelated Scattering (WSSUS) channels, introduced by [6], are the simplest nondegenerate class of processes that exhibit uncorrelated dispersiveness in propagation delay and Doppler shift. These channels are spatial cases of multipath fading channels and carry an uncorrelated scattering property for both time delay and Doppler spread. WSSUS correlation functions, which are given in Equations (5.50) to (5.53), are singular for time delay and Doppler shift variables [17].

$$_b(t,\Delta t;\tau,\eta) = \delta(\eta-\tau)\Psi_b(\Delta t;\tau) \tag{5.50}$$

$$_T(f,f+\Delta f;t,t+\Delta t) = {}_T(\Delta f;\Delta t) \tag{5.51}$$

$$_H(f,f+\Delta f;\beta,\mu) = \delta(\beta-\mu)\Psi_H(\Delta f;\beta) \tag{5.52}$$

$$_x(\tau,\eta;\beta,\mu) = \delta(\eta-\tau)\delta(\beta-\mu)\Psi_x(\tau,\beta) \tag{5.53}$$

Fourier transformations between channel correlation functions are given in Figure 5.8.

The $_b(0;\tau) = {}_b(\tau)$ function is known as the density profile of multipath channels or power delay profile. Average delay is [17]

$$\mu_\tau = \frac{\int_0^{\infty} \tau\ {}_b(\tau)\,d\tau}{_b(\tau)\,d\tau} \tag{5.54}$$

The term

$$\int_0^{\infty} {}_b(\tau)\,d\tau$$

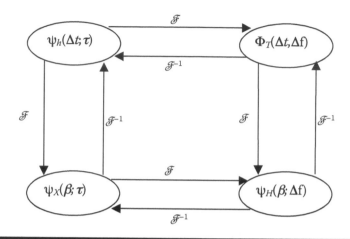

**Figure 5.8 Fourier transformations between WSSUS channel correlation functions.**

is normalized, since $_b(\tau)$ is not a pdf. Then RMS delay spread is

$$\sigma_\tau = \sqrt{\frac{\int_0^\infty (\tau - \mu_\tau)^2 \ _b(\tau)\, d\tau}{\int_0^\infty \ _b(\tau)\, d\tau}} \tag{5.55}$$

The power delay spread is a very important factor for the channel equalizer. If the delay spread exceeds 10 to 20 percent of the symbol duration, then an adaptive equalizer is needed. In typical macro-cell applications, a delay spread is between 1 μs and 10 μs. In micro-cell applications, delay spread is lower. The delay spread changes between 30 and 60 *ns* for indoor mobile communications, where metal and wall structures are effective. In areas where heavy metal construction is a problem, delay spread can reach up to 300 *ns*. The $_T(\Delta t; \Delta f)$ function related to the power delay spectrum is named the space–time, space–frequency correlation function. Function $_T(0; \Delta f) \equiv \ _T(\Delta f)$ defines the frequency correlations of the channels [17]. The compatible bandwidth ($B_c$) corresponds to the minimum $\Delta f$ value of the channel in the case of the correlation coefficient $_T(\Delta f)$. Average delay or average spread is an important measurement for choosing the channel's compatibility bandwidth and therefore, $B_c \alpha (1/\mu_\tau)$ or $B_c \alpha (1/\sigma_\tau)$ relations can be expressed. Function $_H(\beta; 0) \equiv \ _H(\beta)$ is defined as power spectral density (psd) of the Doppler spread. Compatibility time $T_c$ is inverse to the Doppler spread $T_c = 1/B_d$ while $_H(v)$ and $_T(\Delta t)$ functions are considered as the Fourier transform pairs. Calculation of the compatibility time is a vital criterion in evaluating coding performance and intersymbol interference techniques. Doppler spread and consequently compatibility time are directly related to the velocity of the mobile unit. $\Psi_S(\tau,\beta)$ is a scattering function that gives the average power of the channel with time delay $\tau$ and Doppler shift $\beta$.

### 5.3.3 Auto-Correlation of WSSUS Channels

Auto-correlations of the channel outputs also can be defined by using transmission functions. From Equation (5.26) [17],

$$\bar{r}\bar{r}(t,s) = \int_{-\infty}^{\infty}\int_{-\infty}^{\infty} x(t-\tau)\,x^*(s-\eta)\frac{1}{2}E\Big[b(t,\tau)\,b^*(s,\eta)\Big]d\tau d\eta$$

$$= \int_{-\infty}^{\infty}\int_{-\infty}^{\infty} x(t-\tau)\,x^*(s-\eta)\quad {}_b(t,s;\tau,\eta)\,d\tau d\eta \qquad (5.56)$$

Equation (5.56) can be arranged as in [9,17],

$$\bar{r}\bar{r}(t,\Delta t+t) = \int_{-\infty}^{\infty}\int_{-\infty}^{\infty} x(t-\tau)\,x^*(\Delta t+t-\eta)\quad {}_b(\Delta t;\tau)\,\delta(\eta-\tau)\,d\tau d\eta$$

$$= \int_{-\infty}^{\infty} x(t-\tau)\,x^*(t+\Delta t-\tau)\quad {}_b(\Delta t;\tau)\,d\tau \qquad (5.57)$$

At the same time, auto-correlations of the channel outputs can be defined as a scattering function [17].

$$\bar{r}\bar{r}(t,\Delta t+t) = \int_{-\infty}^{\infty}\int_{-\infty}^{\infty} x(t-\tau)\,x^*(\Delta t+t-\tau)\quad {}_s(\tau;\beta)e^{j2\pi v\Delta t}\,d\tau d\beta \qquad (5.58)$$

### 5.3.4 COST 207 Models

In GSM systems, a WSSUS channel is modeled by COST 207 (Cooperation in the Field of Science and Technology, Project #207) and has four different Doppler spectra, as explained in [10,16]. The general model of COST 207 can be given as

$$G(f) = A\exp\left\{-\frac{(f-f_1)^2}{2f_2^2}\right\} \qquad (5.59)$$

COST 207 models are CLASS, GAUS1, GAUS2, and RICE. CLASS is the classical Doppler spectrum where the path delays are less than 500 *ns* ($\tau_i \leq 500$ ns) [10].

$$S_{bb}(f) = \frac{A}{\sqrt{1-(f/f_m)^2}} \qquad |f| \leq f_m \qquad (5.60)$$

GAUS1 is the addition of the two Gaussian functions where path delays are between 500 ns and 2 μs (500 ns $\leq \tau_i \leq$ 2 μs) [10].

$$S_{bb}(f) = G(A, -0.8f_m, 0.05f_m) + G(A_1, 0.4f_m, 0.1f_m) \qquad (5.61)$$

where $A_1$ is 10 dB lower than A.

GAUS2 is the addition of the two Gaussian functions where path delays are greater than 2 μs ($\tau_i \geq$ 2 μs) [10].

$$S_{bb}(f) = G(B, 0.7f_m, 0.1f_m) + G(B_1, -0.4f_m, 0.15f_m) \qquad (5.62)$$

where $B_1$ is 15 dB lower than B.

RICE is the severe case where the classic Doppler spectrum and short paths are combined [10].

$$S_{bb}(f) = \frac{0.41}{2\pi f_m \sqrt{1-(f/f_m)^2}} + 0.91\,\delta(f - 0.7f_m) \qquad |f| \leq f_m \qquad (5.63)$$

The considered channel can be modeled according to various power delay spectrum and path delays. In this case, path scatterings will be 0–7 μs for a typical urban (TU) channel model, 0–10 μs for a bad urban (BU) channel model, 0–0.7 μs for a rural area (RA) channel model, and between 0–2 and 15–20 μs for a hilly terrain (HT) channel model. For a symbol period of 3.7 μs and a carrier frequency of 950 MHz, impulse responses of 50 paths are obtained. The power delay spectrum and impulse responses for TU, BU, RA, and HT channel models are shown in Figures 5.9 to 5.16.

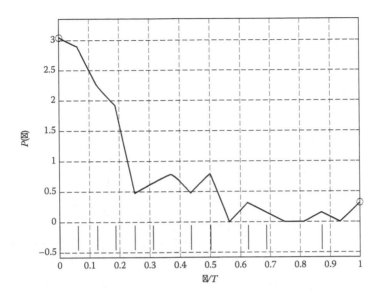

**Figure 5.9** Power delay spectrum of COST 207 TU channel model (Copyright permission from Nobel Press [17]).

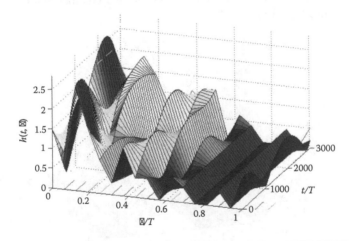

**Figure 5.10** Impulse response of COST 207 TU channel model (Copyright permission from Nobel Press [17]).

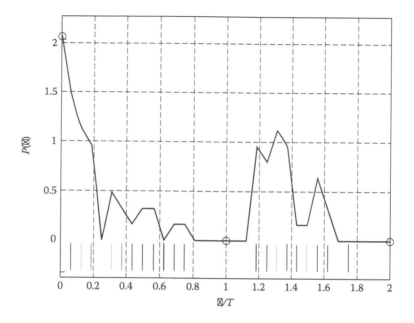

**Figure 5.11 Power delay spectrum of COST 207 BU channel model (Copyright permission from Nobel Press [17]).**

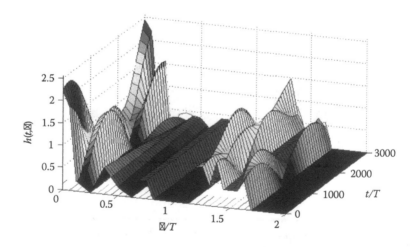

**Figure 5.12 Impulse response of COST 207 BU channel model (Copyright permission from Nobel Press [17]).**

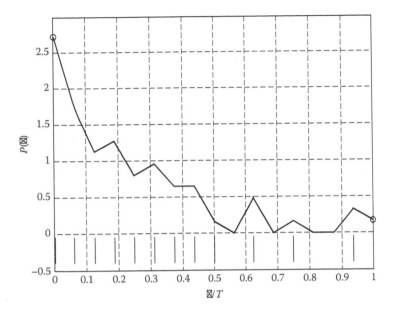

**Figure 5.13    Power delay spectrum of COST 207 RA channel model (Copyright permission from Nobel Press [17]).**

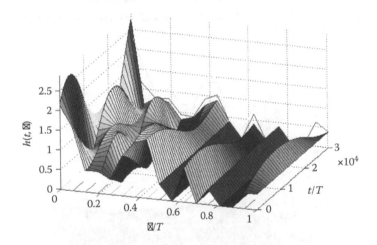

**Figure 5.14    Impulse response of COST 207 RA channel model (Copyright permission from Nobel Press [17]).**

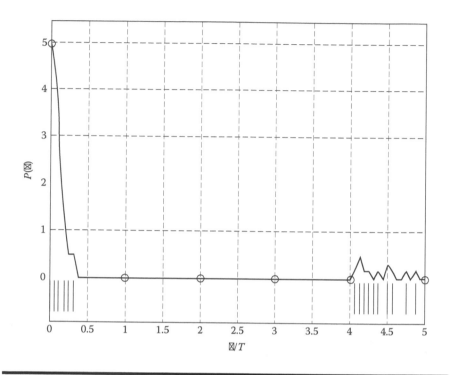

**Figure 5.15    Power delay spectrum of COST 207 HT channel model (Copyright permission from Nobel Press [17]).**

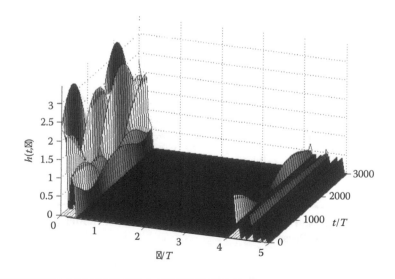

**Figure 5.16    Impulse response of COST 207 HT channel model (Copyright permission from Nobel Press [17]).**

## 5.4 Multiple Input Multiple Output Channels

In wireless communication, multipath fading results in severe amplitude and phase distortion. So it is crucial to combat the effect of the fading at both the remote units and the base stations without additional power or any sacrifice in bandwidth. Space–time coding is a bandwidth and power-efficient method of communication over Rician fading channels that use the benefits of multiple transmit antenna [11]. In MIMO transmissions, high data rates are accomplished by multiple antennas both at the transmitter and the receiver, opening equivalent parallel transmission channels [12,13]. Improved communication reliability results from increased transmit and/or receive diversity [14].

### 5.4.1 MIMO Channel and Space-Time Block Codes

Input of a space–time block encoder is a block of K complex $u_i$, i = 1 ... K, where $u_i$'s are the elements of a higher order modulation constellation, e.g., M-PSK. The space–time block encoder maps the input symbols on entries of a $p \times n_T$ matrix **G**, where $n_T$ is the number of transmit antennas. The entries of the matrix **G** are the K complex symbol $u_i$, their complex conjugate $u_i^*$, and linear combinations of $u_i$ and $u_i^*$. The $p \times n_T$ matrix **G**, which defines the space–time block code, is a complex generalized orthogonal as defined in [15]. This means that the columns of **G** are orthogonal. An example for $n_T = 2$ is the complex generalized orthogonal design

$$G_1 = \begin{bmatrix} g_{11} & \cdots & g_{1n_T} \\ \vdots & & \vdots \\ g_{p1} & \cdots & g_{pn_T} \end{bmatrix} = \begin{bmatrix} u_1 & u_2 \\ -u_2^* & u_1^* \end{bmatrix} \qquad (5.64)$$

Here, consider a simple transmit diversity scheme that improves the signal by simple processing across two transmitter antennas and one/two receiver antenna(s) as shown in Figure 5.17. Assume two modulated signals are simultaneously sent from the two transmitter antennas, $Tx_1$ and $Tx_2$. In

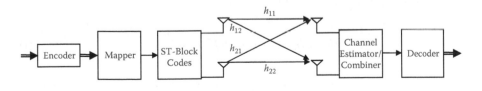

**Figure 5.17  General block diagram of MIMO systems.**

the first coding step, the signal transmitted from the first antenna, $Tx_1$ is denoted as $u_i$ and from the second antenna as $Tx_2$ as $u_{i+1}$. In the second coding step, antenna $Tx_1$ transmits $-u^*_{i+1}$, while antenna $Tx_2$ transmits $u^*_i$ where * means complex conjugate operation. Then assume that a Rician channel with imperfect phase reference fading is constant during these two consecutive symbols. Let the transmitter antenna be $Tx_1$, and the receiver antenna $Rx_1$. The Rician channel between these antennas is defined by $b_{11}$, $b_{12}$, $b_{21}$, and $b_{22}$. The multipath channels $b_{11}$, $b_{12}$, $b_{21}$, and $b_{22}$ are modeled as

$$b_{11}(t) = b_{11}(t + T) = \rho_{11} e^{j\theta_{11}} \tag{5.65a}$$

$$b_{12}(t) = b_{12}(t + T) = \rho_{12} e^{j\theta_{12}} \tag{5.65b}$$

$$b_{21}(t) = b_{21}(t + T) = \rho_{21} e^{j\theta_{21}} \tag{5.65c}$$

$$b_{22}(t) = b_{22}(t + T) = \rho_{22} e^{j\theta_{22}} \tag{5.65d}$$

where $T$ is the symbol duration, $\rho$ is fading amplitude and the term $e^{j\theta}$ is the phase noise.

The received signals are given as

$$r_0 = b_{11}u_1 + b_{21}u_2 + n_0 \tag{5.66a}$$

$$r_1 = -b_{11}u_2^* + b_{21}u_1^* + n_1 \tag{5.66b}$$

$$r_2 = b_{12}u_1 + b_{22}u_2 + n_2 \tag{5.66c}$$

$$r_2 = -b_{12}u_2^* + b_{22}u_1^* + n_3 \tag{5.66d}$$

The noisy symbols can be obtained as

$$\hat{u}_1 = b_{11}^* r_0 + b_{21} r_1^* + b_{12}^* r_2 + b_{22} r_3^* \tag{5.67a}$$

$$\hat{u}_2 = b_{21}^* r_0 - b_{11} r_1^* + b_{22}^* r_2 - b_{12} r_3^* \tag{5.67b}$$

Substituting the appropriate equations we have

$$\hat{u}_1 = (\rho_{11}^2 + \rho_{12}^2 + \rho_{21}^2 + \rho_{22}^2)u_1 + b_{11}^* n_0 + b_{21} n_1^* + b_{12}^* n_2 + b_{22} n_3^* \quad (5.68a)$$

$$\hat{u}_2 = (\rho_{11}^2 + \rho_{12}^2 + \rho_{21}^2 + \rho_{22}^2)u_2 + b_{21}^* n_0 - b_{11} n_1^* + b_{22}^* n_2 - b_{12} n_3^* \quad (5.68b)$$

## 5.4.2 WSSUS MIMO Channels

Consider the discrete-time system model and complex baseband signaling using transmit antennas and receive antennas that are shown in Figure 5.18. The vector symbol $\mathbf{s}[k] = [\mathbf{s}_1[k] \cdots \mathbf{s}_{n_T}[k]]^T$, is composed of the entries of the vector symbols at time $k$. $\mathbf{s}[k]$ is transmitted over the WSSUS MIMO channel denoted as $\mathbf{H}[k]$, where $\mathbf{H}[k]$ is an $n_T \times n_R$ matrix.

The symbols $\mathbf{s}_v[k]$, $1 \leq v \leq n_T$ are drawn from an M-ary alphabet, e.g., a PSK or QAM constellation, with equal transmit power for each antenna [13]. Collecting the received samples in a vector $\mathbf{y}[k] = [\mathbf{y}_1[k], \cdots, \mathbf{y}_{n_R}[k]]$, the input–output relation becomes

$$\mathbf{y}[k] = \mathbf{H}[k]\mathbf{s}[k] + \mathbf{n}[k] \quad (5.69)$$

where $\mathbf{n}[k] = [\mathbf{n}_1[k] \cdots \mathbf{n}_{n_R}[k]]^T$ denotes the complex additive white Gaussian noise (AWGN).

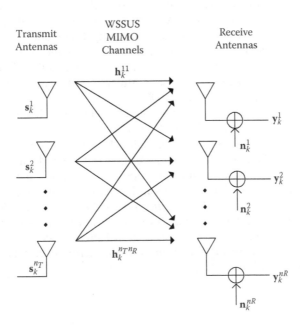

**Figure 5.18** WSSUS MIMO channels.

# References

[1] Proakis, J. G., *Digital Communications,* 4th Edition, New York: McGraw-Hill, 2001.

[2] U•an, O. N., O. Osman, and A. Gumus. Performance of Turbo Coded Signals over Partial Response Fading Channels with Imperfect Phase Reference. *Istanbul University Journal of Electrical & Electronics Engineering* 1(2):149–168, 2001.

[3] U•an, O. N. Trellis Coded Quantization/Modulation over Mobile Satellite Channel with Imperfect Phase Reference. *International Journal of Satellite Communications,* 16:169–175, 1998.

[4] Stavroulakis, P. *Wireless Local Loop: Theory and Applications.* Chichester: John Wiley and Sons Ltd. Publications, 2001.

[5] U•an O. N., O. Osman, and S. Paker. Turbo Coded Signals over Wireless Local Loop Environment. *Int. Journal of Electronics and Commun.* 56(3):163–668, 2002.

[6] Bello, P. A. Characterization of Randomly Time-Variant Linear Channels. *IEEE Transactions on Communication Systems,* CS-11:360–393, Dec. 1963.

[7] Parsons, J. D. *The Mobile Radio Propagation Channel.* New York: Wiley, 1992.

[8] Zadeh, L. A. Frequency Analysis of Variable Networks. *Institute Radio Engineers* 38:291–299, 1950.

[9] Gordon, L. S. *Principles of Mobile Communication,* 2nd Edition. Boston: Kluwer Academic Publishers, 2001.

[10] COST 207. Proposal on Channel Transfer Functions to Be Used in GSM Tests Late 1986. TD(86)51-REV 3 (WG1), September 1986.

[11] Alamouti, S. M. A Simple Transmit Diversity Technique for Wireless Communication. *IEEE Journal of Selected Areas in Comm.* 16(8), October 1998.

[12] Foschini, G., and M. Gans M. On Limits of Wireless Communications in a Fading Environment when Using Multiple Antennas. *Wireless Personal Commun.* 6(3):311–335, Mar. 1998.

[13] Telatar, I. E. Capacity of Multi-Antenna Gaussian Channels. *Eur. Trans. Telecom. (ETT)* 10(6):585–596, Nov. 1999.

[14] Wittneben, A. Base Station Modulation Diversity for Digital SIMUL-CAST. *Proc. IEEE Vehicular Technology Conf. (VTC)*:505–511, May 1991.

[15] Tarokh, V., N. Seshadri, and A. Calderbank A. Space-Time Codes for High Data Rate Wireless Communication: Performance Criterion and Code Construction. *IEEE Transaction on Information Theory* 44:744–765, March 1998.

[16] U•an, O. N., and O. Osman. *Communication Theory and Engineering Applications* (in Turkish). Istanbul: Nobel Press, 2006.

[17] Stüber, G. L. *Principles of MobileCommunication,* 2nd Edition. Kluwer Academic Publisher, 2001.

# Chapter 6

# Channel Equalization

In wireless communications, adaptive filters are used for channel equalization and/or estimation of channel coefficients. Adaptive filters are digital filters capable of self adjustment. Digital filters are typically a special type of finite impulse response (FIR) and infinite impulse response (IIR) filters. Adaptive filters can also be separated in two main groups; semi-blind and blind filters, according to their information knowledge of parameters to be estimated by filtering [1–9].

## 6.1 Semi-Blind Equalization with Adaptive Filters

In semi-blind filters, some of the coefficients are assumed to be known during estimation. The main semi-blind filters are least mean square (LMS), recursive least squares (RLS), Kalman and genetic algorithm (GA).

### 6.1.1 LMS Algorithm

The LMS algorithm is an adaptive algorithm based on the gradient-based method of steepest decent. The LMS algorithm uses the estimates of the gradient vector from the available data. LMS incorporates an iterative procedure that makes successive corrections to the weight vector in the direction of the negative of the gradient vector, which eventually leads to the minimum mean square error. The LMS algorithm is relatively simple, compared to the other approaches. It does not require correlation function calculation nor does it require matrix inversions.

In an adaptive FIR filter, the estimated output signal $\hat{d}(n)$ is compared with the input $d(n)$ to produce the error signal $e(n)$. The estimated signal is expressed as

$$\hat{d}(n) = \sum_{m=0}^{M-1} b_m(n)x(n-m) = \mathbf{h}(n)^{\mathbf{T}}\mathbf{x}(n) \qquad (6.1)$$

where

$$\mathbf{x}(n) = \begin{bmatrix} x(n) \\ x(n-1) \\ x(n-2) \\ \vdots \\ x(n-M+1) \end{bmatrix} \quad \text{and} \quad \mathbf{h}(n) = \begin{bmatrix} b_0(n) \\ b_1(n) \\ b_2(n) \\ \vdots \\ b_{M-1}(n) \end{bmatrix}$$

The Jacobian expression $J_n$ can be written as

$$J_n = e^2(n) = \left[ d(n) - \mathbf{h}(n)^{\mathbf{T}}\mathbf{x}(n) \right]^2 \qquad (6.2)$$

Equation (6.2) shows the squared error between desired and estimated signals. To minimize error, derivation of $J_n$ with respect to $\mathbf{h}(n)$ can be taken as

$$\frac{J_n}{\mathbf{h}(n)} = -2e(n)\mathbf{x}(n) \qquad (6.3)$$

The weight vector equation of the method of steepest descent is given as

$$\mathbf{h}(n+1) = \mathbf{h}(n) + \mu\left( \frac{J_n}{\mathbf{h}(n)} \right) \qquad (6.4)$$

where $\mu$ is the step-size parameter and controls the convergence characteristics of LMS algorithm. If $\mu$ is very small then the algorithm converges very slowly. A large value of $\mu$ may lead to a faster convergence but may be less stable around the minimum value. Applying Equation (6.3) to Equation (6.4),

$$\mathbf{h}(n+1) = \mathbf{h}(n) + 2\mu \; e(n)\mathbf{x}(n) \tag{6.5}$$

In the method of steepest descent, the foremost problem is the computation involved in finding the values in real time. The LMS algorithm on the other hand simplifies this by using instantaneous values instead of actual values. The LMS algorithm is initiated with an arbitrary value $\mathbf{h}(0)$ for the weight vector at $n = 0$. The successive corrections of the weight vector eventually lead to the minimum value of the mean squared error. The main problem of LMS algorithm is that it takes a long time for the filter coefficients $\mathbf{h}(n)$ to converge because they are adjusted at an identical $\mu$ rate. Recursive least squares algorithm or Kalman filter improves this disadvantage at the expense of an increment of computing complexity.

## 6.1.2 Recursive Least Squares Algorithm and Kalman Filtering

The recursive least squares algorithm is used to find the filter coefficients that relate to producing the recursively least squares of the error signal. In RLS filters, the aim is to minimize a weighted least squares error function. The Jacobian expression in Equation (6.2) can be rewritten employing a factor $b$ denoted as the forgetting factor [1–2].

$$J_n = \sum_{i=0}^{n} b^{n-i} e^2(i) = \sum_{i=0}^{n} b^{n-i} \left[ d(i) - \mathbf{x}(i)\mathbf{h}(n) \right]^2 \tag{6.6}$$

The forgetting factor is in the range of {0,1}. To minimize Equation (6.6), we derive the value of $\mathbf{h}(n)$ as

$$\frac{J_n}{\mathbf{h}(n)} = -2 \sum_{i=0}^{n} b^{n-i} \left[ -\mathbf{x}(i) \right] \left[ d(i) - \mathbf{x}(i)^T \mathbf{h}(n) \right] = 0 \tag{6.7}$$

In vector form, representation of the right side of Equation (6.7) is as

$$\mathbf{R}(n)\mathbf{h}(n) = \mathbf{p}(n) \tag{6.8}$$

Then, $\mathbf{h}(n)$ is simply

$$\mathbf{h}(n) = \mathbf{R}(n)^{-1}\mathbf{p}(n) \tag{6.9}$$

The recursive formulas of Equations (6.8) and (6.9) are

$$\mathbf{R}(n) = b\mathbf{R}(n-1) + \mathbf{x}(n)\mathbf{x}(n)^{T} \tag{6.10}$$

$$\mathbf{p}(n) = b\mathbf{p}(n-1) + d(n)\mathbf{x}(n) \tag{6.11}$$

Then inverse matrix of $\mathbf{R}(n)^{-1}$ can be stated as

$$\mathbf{R}(n)^{-1} = \frac{1}{b}\left[\mathbf{R}(n-1)^{-1} - \frac{\mathbf{R}(n-1)^{-1}\mathbf{x}(n)\mathbf{x}(n)^{T}\mathbf{R}(n-1)^{-1}}{b + \mathbf{x}(n)^{T}\mathbf{R}(n-1)^{-1}\mathbf{x}(n)}\right] \tag{6.12}$$

Applying Equation (6.12) to (6.9),

$$\mathbf{h}(n) = \mathbf{R}(n)^{-1}\mathbf{p}(n)$$
$$= \mathbf{h}(n-1) + \mathbf{k}(n)\left[e(n|n-1)\right] \tag{6.13}$$

where $\mathbf{k}(n)$ is Kalman gain and is calculated as,

$$\mathbf{k}(n) = \left[\frac{\mathbf{R}(n-1)^{-1}\mathbf{x}(n)}{b + \mathbf{x}(n)^{T}\mathbf{R}(n-1)^{-1}\mathbf{x}(n)}\right]$$

and tentative estimation error $e(n|n-1)$ is as,

$$e(n|n-1) = d(n) - \mathbf{h}(n-1)^{T}\mathbf{x}(n)$$

In the RLS algorithm, tentative error is based on the correct estimation of $\mathbf{h}(n-1)$.

## 6.1.3 Genetic Algorithms (GAs)

Genetic algorithms (GAs) are search and optimization algorithms based on the principle of natural evolution and population genetics. GAs were first proposed by Holland [10] and have been applied successfully to many

engineering and optimization problems. GAs use different genetic operators to manipulate individuals in a population over several generations to
gradually improve their fitness. GAs maintain a set of candidate solutions
called, a population. Candidate solutions are usually represented as strings
called chromosomes, coded with a binary character set {0,1}. A GA has a
population of individuals. Each individual represents a potential solution
to the problem in question. Individuals are evaluated to give a measure
of their fitness.

To involve the best solution candidate (or chromosome), the GA
employs the genetic operators of selection, crossover and mutation for
manipulating the chromosomes in a population. The flow chart of a simple
GA is shown in Figure 6.1.

In the selection step, individuals are chosen to participate in the
reproduction of new individuals. Different selection methods, such as
stochastic selection or ranking-based selection, can be used. The crossover
operator combines characteristics of two parent individuals to form two
offspring. Mutation is a random alteration with a small probability of the
binary value of a string position. This operation will prevent a GA from
being trapped in a local minimum. The fitness evaluation unit in the flow
chart acts as an interface between the GA and the optimization problem.
Information generated by this unit about the quality of different solutions
is used by the selection operation in the GA. More details about GAs can
be found in [11,15].

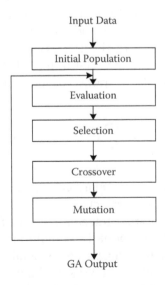

**Figure 6.1   Flow chart of a genetic algorithm.**

## 6.1.4 WSSUS MIMO Channel Equalization Using GA

The main effects that cause and influence multipath propagation, are reflection, scattering, shadowing, diffraction, and path loss. The multipath propagation results in frequency and time-selective fading of the received signal at the receiver. In digital communication, systems require the knowledge of the transmission channel's impulse response. Because this knowledge usually is not available, the problem of channel estimation arises. From the point of view of systems theory, channel estimation is a particular form of (linear) system identification that, in our case, is complicated by three main properties of the radio channel: (1) it consists of multiple propagation paths and is therefore frequency-selective, (2) its discrete-time equivalent baseband impulse response may be mixed-phase and (3) in a mobile environment, it is time-variant [14]. Because of the movement of the mobile station and hence the changing channel characteristics, the channel has to be treated as a time variant, multipath fading channel. This can be characterized by statistical channel models such as WSSUS (wide-sense stationary uncorrelated scattering) models. WSSUS models require only two sets of parameters to characterize fading and multipath effects: the power delay profiles and Doppler power spectra. Therefore, these can efficiently be used within software (and hardware) simulation systems. Both parameter sets can be described by a single function called scattering function.

In a mobile scenario, the physical multipath radio channel is time-variant with a baseband impulse response depending on the time difference $\tau$ between the observation and excitation instants as well as the (absolute) observation time $t$. The stochastic zero means the Gaussian stationary uncorrelated scattering (GSUS) model should be adopted, leading to the following impulse response of the composite channel [15]

$$b^c(\tau, t) = \frac{1}{\sqrt{N_e}} \sum_{v=1}^{N_e} e^{j(2\pi f_{d,v} t + \Theta_v)} \cdot g_{TR}(\tau - \tau_v) \qquad (6.14)$$

where $N_e$ is the number of elementary echo paths, $f_{d,r}$ is Doppler frequencies, $\Theta_v$ is initial phases and $\tau_v$ is echo delay times from random variables. Here, $g_{TR}(\tau)$ denotes combined transmit/receiver filter impulse response, and the subscript in $b^c(\cdot)$ suggests its continuous-time property.

During transmission, the signals are generally corrupted due to the severe transmission conditions of WSSUS channels. So it is necessary to minimize these effects and receive the signals with minimum errors by using equalizers and decoders together. At the genetic-based channel estimator, $\mathbf{y}(k)$ is the output vector at time $k$. The GA will be used to search for channel coefficients $\mathbf{h}(k)$ that will minimize the square root of error between the output vector $\mathbf{y}(k)$ and $\mathbf{s}(k) * \mathbf{h}(k)$ at time $k$.

The following performance index can be defined.

$$\mathbf{y}(k) = \mathbf{s}(k) * \mathbf{h}(k) + \mathbf{n}(k) \tag{6.15}$$

$$\mathbf{J}(k) = \sqrt{\mathbf{y}(k) - \mathbf{s}(k) * \mathbf{h}(k)} \tag{6.16}$$

The genetic-based channel estimator searches for the optimal value of channel coefficients $\mathbf{h}(k)$ that will minimize $\mathbf{J}$. The following steps describe the operation of the proposed GA-based estimator.

At time step $k$:

1. Evaluate $\mathbf{y}(k)$ using the WSSUS multipath channel model.
2. Use GA search to find $\mathbf{h}(k)$ which will minimize the performance index $\mathbf{J}$. The WSSUS channel model should be used to find $\mathbf{y}(k)$ for different values of $\mathbf{h}(k)$.
3. Apply the optimal value of $\mathbf{h}(k)$ generated in step 2 to $\mathbf{y}(k)$.
4. Repeat for time $k + 1$.

## 6.2 Blind Equalization with Adaptive Filters

If both the received sequence $r(k)$ and some transmitted data $u(k)$ (training sequence) are given the minimum mean square error (MMSE-$(\ell,k_0)$) equalizer coefficients can be calculated using the well-known normal Equation (6.17) [11]

$$\mathbf{e}_{MMSE(k_0)} = \mathbf{R}_{rr}^{-1} \mathbf{r}_{ru} \tag{6.17}$$

with

$$\mathbf{r}_{ru} \quad E\left\{ \mathbf{r}_k \, u(k - k_0) \right\} \tag{6.18}$$

$$\mathbf{R}_{uu} \quad E\left\{ \mathbf{u}_k \mathbf{u}_k^* \right\} \tag{6.19}$$

where $\mathbf{r}_{ru}$ and $\mathbf{R}_{uu}$ denote the cross-correlation vector and the non-singular $(l + 1) \times (l + 1)$ Hermitian Toeplitz autocorrelation matrix, respectively, and the vectors $r_k$ and conjugate transpose form $r_k^*$ are defined as

$$\mathbf{r}_k \quad \left[ r^*(k), r^*(k-1), \quad , r^*(k-l) \right]^T \tag{6.20}$$

$$\mathbf{r}_k^* \quad \left[ r(k), r(k-1), \quad , r(k-l) \right] \tag{6.21}$$

Then the MMSE-$(\ell, k_0)$ equalizer is

$$\mathbf{e}_{MMSE(k_0)} \quad \left[ e_{MMSE}(o), \quad , e_{MMSE}( ) \right]^T \tag{6.22}$$

In the noiseless case, it approximates the channel's inverse system (deconvolution, zero forcing) in order to minimize intersymbol interference (ISI). If additive noise is present, however, its coefficients adjust differently to minimize the total mean squared error (MSE) in the equalized sequence due to ISI and noise.

The fundamental purpose of blind channel estimation is to derive the channel characteristics from the received signal only, i.e., without access to the channel input signal by means of training sequences [16–23]. Depending on the different ways to extract information from the received signal, some classes of algorithms such as eigenvector algorithm (EVA) can be used [16]. Figure 6.2 shows a block diagram of the EVA equalizer. The transmitted data $u_k$ is an independent, identically distributed (i.i.d.) sequence of random variables with zero mean, variance $\sigma_u^2$, skewness $\gamma_3^u$, and kurtosis $\gamma_4^u$ [11]. Each symbol period $T$, $u_k$ takes a (possibly complex) value from a finite set. For this reason, the channel input random process clearly is non-Gaussian with a non-zero kurtosis ($\gamma_4^u$ 0), while its skewness vanishes ($\gamma_3^u = 0$) due to the even probability density function of typical digital modulation signals, such as phase shift keying (PSK), quadrature amplitude modulation (QAM), and so forth. The objective is

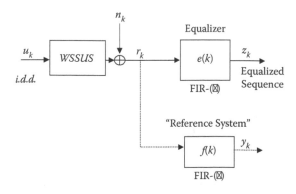

**Figure 6.2  Block diagram of EVA equalizer.**

to determine the MMSE-$(\ell,k_o)$ equalizer coefficients without access to the transmitted data, i.e., from the received sequence $r_k$ only.

Similar to Shalvi and Weinstein's maximum kurtosis criterion [14,15], the EVA solution to blind equalization is based on a maximum "cross-kurtosis" quality function [16]:

$$c_4^{zy}(0,0,0) = E\left\{\left|z_k\right|^2\left|y_k\right|^2\right\} - E\left\{\left|z_k\right|^2\right\}E\left\{\left|y_k\right|^2\right\}$$
$$-\left|E\left\{z_k^* y_k\right\}\right|^2 - \left|E\left\{z_k y_k\right\}\right|^2 \tag{6.23}$$

i.e., the cross-kurtosis between $u_k$ and $r_k$, can be used as a measure for equalization quality:

$$\max \left| c_4^{zy}(0,0,0) \right| = \begin{cases} subject \quad to \\ r_{zz}(0) = \sigma_u^2 \end{cases} \tag{6.24}$$

The equalizer output $z_k$, can easily be expressed in terms of the equalizer input $r_k$ by replacing $z_k$ in Equation (6.23) with

$$z_k = r_k * e(k) = \mathbf{r}_k^* \mathbf{e} \tag{6.25}$$

where "*" denotes the convolution operator. In this way, from (6.24)

$$\max \left| \mathbf{e}^* \mathbf{C}_4^{yr} \mathbf{e} \right| = \begin{cases} subject \ to \\ \mathbf{e}^* \mathbf{R}_{rr} \mathbf{e} = \sigma_u^2 \end{cases} \tag{6.26}$$

where the Hermitian $(l + 1) \times (l + 1)$ cross-cumulant matrix

$$\mathbf{C}_4^{yr} \quad E\left\{\left|y_k\right|^2 \mathbf{r}_k \mathbf{r}_k^*\right\}$$
$$- E\left\{y_k^* \mathbf{r}_k\right\}E\left\{y_k \mathbf{r}_k^*\right\}$$
$$- E\left\{y_k \mathbf{r}_k\right\}E\left\{y_k^* \mathbf{r}_k^*\right\} \tag{6.27}$$
$$- E\left\{\left|y_k\right|^2\right\}E\left\{\mathbf{r}_k \mathbf{r}_k^*\right\}$$

contains the fourth order cross-cumulant

$$\left[\mathbf{C}_4^{yr}\right]_{i_1,i_2} = c_4^{yr}(-i_1,0,-i_2) \tag{6.28}$$

in row $i_1$ and column $i_2$ with $i_1, i_2 \in (0, 1, \cdots, \Box]$. The quality function (6.26) is quadratic in the equalizer coefficients. Its optimization leads to a closed-form expression in the guise of the generalized eigenvector problem [16]

$$\mathbf{C}_4^{yr} \mathbf{e}_{EVA} = \lambda \mathbf{R}_{rr} \mathbf{e}_{EVA} \tag{6.29}$$

which we term an "EVA equation." The coefficient vector

$$\mathbf{e}_{EVA} \quad \left[ e_{EVA}(0), \quad , e_{EVA}(\ ) \right]^T \tag{6.30}$$

is obtained by choosing the eigenvector of $\mathbf{R}_{rr}^{-1} \mathbf{C}_4^{yr}$ associated with the maximum magnitude eigenvalue $\lambda$ is called the "EVA-($\ell$) solution" to the problem of blind equalization.

## 6.2.1 ML Channel Equalization

At the $i$th epoch, the m-bit information vector $\mathbf{d}(i) = (d_1,(i), d_2(i), \ldots, d_m(i))^T$ is fed to a transmitter, followed by m/(m + 1) convolutional encoder and MPSK mappers [28]. The output of this system is denoted as $g(\mathbf{d}(i))$, where an inverse FFT operation is applied to decrease the fading effect of the selective channel. If we consider the transmission of coded signals over a nondispersive, narrow-band fading channel, the received signal has the form of

$$y(i) = \rho(i) g(\mathbf{d}(i)) e^{j\theta(i)} + n(i) \tag{6.31}$$

where $x(i) = g(\mathbf{d}(i))$. The following assumptions are imposed on the received signal model:

> **AS1**: $n(i)$ is a zero mean, i.i.d., AWGN sequence with the variance of $\sigma$.
> **AS2**: $\mathbf{d}(i)$ is an $m$ digit binary, equiprobable input information sequence of coding scheme.
> **AS3**: $\rho(i)$ represents a normalized fading magnitude having Rician probability density function (p.d.f.), and $\theta(i)$ represents the phase distortion due to the effect of Doppler shift at the $i$th signaling interval.
> **AS4**: Slowly varying nondispersive fading channels are considered.

The basic problem addressed here is the estimation of the Rician channel parameters from a set of observations. Due to its asymptotic efficiency, the focus should be on the ML approach to the problem at hand. Based on the assumption **AS1**, the maximum likelihood metric for **d** is proportional to the conditional probability density function of the received data sequence of length $N$ (given **d**) and is of the form

$$m_l(\rho, \theta, d) = \sum_{i=1}^{N} \left| y(i) - \rho_l g(\mathbf{d(i)}) e^{j\theta_l} \right|^2 \tag{6.32}$$

where **d** = $(d(1),\ d(2),\ \dots\ ,\ d(N))$. Because the transmitted data sequence is not available at the receiver, the ML metric to be maximized can be obtained after averaging $m_l(g, d)$ over all possible transmitted sequences $d$,

$$m_l(g) = \sum_d m_l(g, d)$$

Then the ML estimator of $\lambda = [\rho_l, \theta_l, \sigma_l^2]$ is the one that maximizes the average ML metric

$$\hat{\lambda} = \arg\ \max_{\lambda}[m_l(\rho, \theta)]$$

Here slowly varying fading channels are assumed, $\rho$ and $\theta$ remain constant during the $l$ observation interval. The metric for **d** can be obtained by averaging Equation (6.32) over all possible values of **d**$(i)$,

$$m_l(\rho, \theta) = \sum_d \sum_{i=1}^{N} \left| y(i) - \rho_l g(\mathbf{d}(i)) e^{j\theta_l} \right|^2 \tag{6.33}$$

Because the direct minimization of Equation (6.33) requires the evaluation of Equation (6.32) over all possible transmitted sequences, this increases the complexity exponentially with the data length $N$. Fortunately, a general iterative approach to compute ML estimates known as the Baum–Welch algorithm [23–28] can be used to significantly reduce the computational burden of an exhaustive search.

## 6.2.2  Baum–Welch Channel Equalization

The Baum–Welch algorithm is an iterative ML estimation procedure originally developed for estimating the parameters of a probabilistic function of a Markov chain [23]. It is based on the maximization of an auxiliary function related to the Kullback–Leibler information measure instead of the likelihood function. It has found applications in diverse parameter estimation problems, especially where there are multiple parameters and highly nonlinear likelihood functions. The Kullback–Leibler information measure is defined as

$$KL(\lambda_1, \lambda_2) = E_{\lambda_1} \log\{\frac{f_{\lambda_1}(y)}{f_{\lambda_2}(y)}\} = E_{\lambda_1}\{\log f_{\lambda_1}(y)\} - E_{\lambda_1}\{\log f_{\lambda_2}(y)\} \quad (6.34)$$

where $f_\lambda(y)$ is the family of distribution identical to the unknown parameter of the available data. Kullback–Leibler gives some measure of discrepancy between $f_{\lambda_1}(y)$ and $f_{\lambda_2}(y)$.

Minimization of $KL(\lambda_1, \lambda_2)$ with respect to $\lambda_2$ is equivalent to maximizing the second term of Equation (6.34), because the term $E_{\lambda_1}\{\log f_{\lambda_1}(y)\}$ is fixed. The auxiliary function used in BW algorithm is therefore defined as an approximation to the second term of Equation (6.34) as

$$\Theta(\lambda_1, \lambda_2) \equiv \sum_\phi f_{\lambda_1}(y, \varphi) \log(f_{\lambda_2}(y, \varphi)) \quad (6.35)$$

where

$$\varphi = \{\varphi(i)\}_{i=1}^{N}$$

takes all possible state sequences. The following theorem, which forms the basis for the BW algorithm, explains the reason for the choice of $\Theta(\lambda_1, \lambda_2)$ in Equation (6.35).

Theorem: $\Theta(\lambda, \lambda') \geq \Theta(\lambda, \lambda)$ implies $f_\lambda(y) \geq f_\lambda(y)$

The proof of the theorem is given by Baum et al. [23]. The iterative BW algorithm for the evaluation of the estimates starts with an initial guess $\lambda^{(0)}$. At the $(t + 1)th$ iteration, it takes estimate $\hat{\lambda}^{(t)}$ from the $t$th iteration as an initial value and reestimates $\hat{\lambda}^{(t+1)}$ by maximizing $\Theta(\hat{\lambda}^{(t)}, \lambda')$ over $\lambda'$.

$$\hat{\lambda}^{(t+1)} = \arg\min_{\lambda'} \Theta(\hat{\lambda}^{(t)}, \lambda') \quad (6.36)$$

This procedure is repeated until the parameters converge to a stable point. It can be shown that the explicit form of the auxiliary function at the $i$th observation interval has the form

$$\Theta(\hat{\lambda}^{(t)}, \lambda'') = C + \sum_{i=1}^{N} \sum_{m=1}^{P} \gamma_m^{(t)}(i) \left\{ \frac{1}{\sigma_i'^2} \left| y(i) - p_i' g(\varphi_m) e^{j\theta'_i} \right|^2 \right\} \quad (6.37)$$

where $\gamma_m^{(t)}(i) = f_{\lambda'_t}(y, \varphi(i) = \varphi_k)$ is the joint likelihood of the data $y$ and being at state $\varphi_k$ at time $i$. Using the definitions of forward $\alpha_k(i)$ and backward $\beta_k(i)$ variables,

$$\beta_j^{(t)}(i) = \sum_{j=1}^{F} a_{m,j} \beta_j^{(t)}(i+1) f_j(y(i+1)) \quad i = N-2,\dots,0 \quad (6.38)$$

$$\alpha_m^{(t)}(i) = \sum_{m=1}^{F} a_{m,j} \alpha_m^{(t)}(i-1) f_j(y(t)) \quad i = 1,\dots,N-1 \quad (6.39)$$

Here $a_{mj}$ is the state transition probability. $f_j(y)$ is the probability density function of the observation and has a Gaussian nature:

$$f_k(y(i)) = \frac{1}{\sigma_i \sqrt{2\pi}} \exp\left\{ -\frac{1}{2\sigma_i^2} \left\| y(i) - \eta_k \right\|^2 \right\} \quad (6.40)$$

$\eta_k = p_i g_z(\phi_k)$ is the probable observed signal and $\sigma_i^2$ is the variance of the Gaussian noise. The initial conditions of $\beta$ and $\alpha$ are

$$\beta_j^{(t)}(N-1) = 1 \qquad 1 \leq j \leq F \quad (6.41)$$

$$\alpha_m^{(t)}(0) = p_m(y(0)) \quad 1 \leq m \leq F \quad (6.42)$$

$p_m$ is the initial probability of the signals. The probabilities of the partial observation sequences $\gamma_k(i)$ can be recursively obtained as [23–26], given the state $\phi_k$ at time $i$:

$$\gamma_k^{(t)}(i) = \alpha_k^{(t)}(i) \beta_k^{(t)}(i) \quad (6.43)$$

The iterative estimation formulas can be obtained by setting the gradient of the auxiliary function to zero $\nabla_{\rho_l}\Theta = 0$, $\nabla_{\theta_l}\Theta = 0$, $\nabla_{\sigma_l^{\prime 2}}\Theta = 0$. Then,

$$\nabla_{\rho_l}\Theta = \frac{2}{\sigma_l^{\prime 2}}\sum_{i=1}^{N}\sum_{m=1}^{P}\gamma_m^{(t)}(i)\ \left\{\left[y(i)-\rho_l'g(\varphi_m)\right]\dot{g}(\varphi_m)\right\} \qquad (6.44)$$

$$\nabla_{\theta_l}\Theta = \frac{2}{\sigma_l^{\prime 2}}\sum_{i=1}^{N}\sum_{m=1}^{P}\gamma_m^{(t)}(i)$$

$$\cdot\ \left\{\begin{array}{l}\left[r(i)-\rho_l'e^{j\theta_l'}(g_1(\varphi_m)+..+g_{n_T}(\varphi_m))\right]\\[2mm] je^{-j\theta_l'}\rho_l'(g_1(\varphi_m)+..+g_{n_T}(\varphi_m))^*\end{array}\right\} \qquad (6.45)$$

$$\nabla_{\sigma_l^{\prime 2}}\Theta = \frac{N}{2\sigma_l^{\prime 2}}+\frac{N}{2\sigma_l^{\prime 4}}\sum_{i=1}^{N}\sum_{m=1}^{P}\gamma_m^{(t)}(i)\ \left\{\left[y(i)-\rho_l'g(\varphi_m)\right]^2\right\} \qquad (6.46)$$

Based on these results, the proposed algorithm can be summarized as

**Initialization:** Set the parameters to some initial value $^{(0)} = (\rho_l^{(0)}, \theta_l^{(0)}, \sigma_l^{2(0)})$.
**Step 1:** Compute the forward and backward variables from Equation (6.43) to obtain $\gamma_m^{(t)}(i)$.
**Step 2:** Obtain $\hat{\rho}_l^{(t+1)}$, $\theta_l^{(0)}$ by solving the set of linear equations provided by Equations (6.44) and (6.45).
**Step 3:** Solve $\nabla_{\sigma_l^{\prime 2}}\Theta = 0$ from Equation (6.46) to obtain $\sigma_l^{\prime 2(t+1)}$.
**Step 4:** Repeat Step 1 to Step 3 until $\left\| \prime^{(t+1)} - \prime^{(t)}\right\| \leq \mu$, where $\mu$ is a predefined tolerance parameter.
**Step 5:** Employ a decoding algorithm on $\gamma^{(t)}$ to recover the transmitted symbols.

# References

[1] Stüber, G. L. *Principles of Mobile Communications*. Norwell, MA: Kluwer Academic Publishers, 2001.
[2] Buyukatak, K. Evaluation of 2-D Images for Turbo Codes. Istanbul Technical University, Ph.D. Thesis, 2004.

[3] U•an, O.N., K. Buyukatak, E. Gose, O. Osman, and N. Odabasioglu, Performance of Multilevel-Turbo Codes with Blind/Non-blind Equalization over WSSUS Multipath Channels. *Int. J. of Commun. Syst.* 19:281–297, 2006.

[4] Osman, O. Blind Equalization of Multilevel Turbo Coded-Continuous Phase Frequency Shift Keying over MIMO Channels. *Int. J. of Communication Systems* 20(1):103–119, 2007.

[5] Bello, P. A. Characterization of Randomly Time-Variant Linear Channels. *IEEE Transactions on Communication Systems* CS-11:360–393, Dec. 1963.

[6] Boss, D., B. Jelonnek, and K. Kammeyer. Eigenvector Algorithm for Blind MA System Identification. *Elsevier Signal Processing* 66(1):1–26, April 1998.

[7] Jelonnek, B., D. Boss, and K. D. Kammeyer. Generalized Eigenvector Algorithm for Blind Equalization. *Elsevier Signal Processing* 61(3):237–264, September 1997.

[8] Jelonnek, B., and K. D. Kammeyer. A Blind Adaptive Equalizer based on Lattice/All Pass Configuration. In *Proc. EUSIPCO-92* II:1109–1112, August 1992.

[9] Haykin, S. *Adaptive Filter Theory,* 3rd Edition. Upper Saddle River, NJ: Prentice Hall, 1996.

[10] Holland, J. *Adaptation in Natural and Arti"cial Systems.* Cambridge, MA: MIT Press, 1975.

[11] Bello, P. A. Characterization of Randomly Time-Variant Linear Channels. *IEEE Transactions on Communication Systems* CS-11:360–393, Dec. 1963.

[12] U•an, O. N., and O. Osman. Turbo Trellis Coded Modulation (TTCM) with Imperfect Phase Reference. *Istanbul University-Journal of Electrical and Electronics Engineering (IU-JEEE)* 1:1–5, 2001.

[13] U•an, O. N., O. Osman, and S. Paker S. Turbo Coded Signals over Wireless Local Loop Environment. *International Journal of Electronics and Communications (AEU)* 56(3):163–168, 2002.

[14] Gose, E., H. Pastaci, O. N. U•an, K. Buyukatak, and O. Osman O. Performance of Transmit Diversity-Turbo Trellis Coded Modulation (TD-TTCM) over Genetically Estimated WSSUS MIMO Channels. *Frequenz* 58:249–255, 2004.

[15] Gose, E. Turbo Coding over Time Varying Channels. Ph.D. Thesis, Yildiz Technical University, 2005.

[16] Telatar, I. E. Capacity of Multi-Antenna Gaussian Channels. *Eur. Trans. Telecom.(ETT)* 10:6:585–596, Nov. 1999.

[17] Wittneben, A. Base Station Modulation Diversity for Digital SIMUL-CAST. In *Proc. IEEE Vehicular Technology Conf. (VTC)*:505–511, May 1991.

[18] Hoeher, P. A Statistical Discrete-Time Model for the WSSUS Multi Path Channel. *IEEE Trans. on Vehicular Technology* VT-41(4):461–468, April 1992.

[19] COST 207. *Digital Land Mobile Radio Communications.* Office for Official Publications of the European Communities, Abschlussbericht, Luxemburg, 1989.

[20] Goldberg, D. E. *Genetic Algorithms in Search, Optimization, and Machine-Learning*. Reading, MA: Addison-Wesley, 1989.

[21] Osman, O. Performance of Turbo Coding. Ph.D. Thesis, Istanbul University, 2003.

[22] U•an, O. N. Performance Analysis of Quadrature Partial Response-Trelllis Coded Modulation (QPR-TCM). Ph.D. Thesis, Istanbul Technical University, 1994.

[23] Baum, L. E., T. Petrie, G. Soules, N. and Weiss. Maximization Technique Occurring in the Statistical Analysis of Probabilistic Functions of Markov Chains. *The Annals of Mathematical Statistics* 41:164–171, Jan. 1970.

[24] Erkurt, M., and J. G. Proakis. Joint Data Detection and Channel Estimation for Rapidly Fading Channels. In *IEEE Globecom*:910–914, 1992.

[25] Rabiner, L. R. A Tutorial on Hidden Markov Models and Selected Applications in Speech Recognition. *Proceeding of IEEE* 77:257–274, 1989.

[26] Kaleh, K., and R. Vallet. Joint Parameter Estimation and Symbol Detection for Linear or Nonlinear Unknown Channels, *IEEE Trans. on Commun.* 42:2406–2413, July 1994.

[27] Foschini, G.J., and M. J. Gans MJ. On Limits of Wireless Communications in a Fading Environment when Using Multiple Antennas. *Wireless Personal Communications* 6:311–335, 1998.

[28] Cirpan, H. A., and O. N. U•an. Blind Equalization of Quadrature Partial Response-Trellis Coded Modulated Signals in Rician Fading. *International Journal of Satellite Communications* 19:159–168, March–April, 2001.

# Chapter 7

# Turbo Codes

Since 1993, the date of the introduction of turbo-codes, these codes have received great attention and found practical applications [1]. They have reasonable complexity, powerful error correcting capability, and the ability to change in terms of adaptation of different block lengths and code rates. In a classic turbo structure, there are two parallel convolutional encoders. First, input data is interleaved before entering the second encoder as in Figure 7.1. At the puncturing stage a code word is constructed consisting of input bits—called systematic—followed by the parity check bits from the first encoder and then the parity bits from the second encoder.

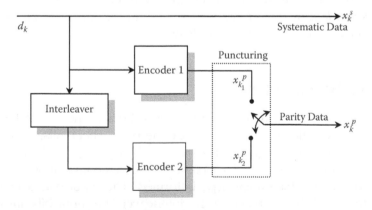

**Figure 7.1   General structure of turbo encoder.**

There may be multiple turbo encoders with more than two branches. The convolutional code at every branch is called the constituent code (CC). The CCs can have similar or different generator functions. Here, the usual configuration with two branches having the same CC will be explained. As an example, consider a turbo-code (TC) encoder with an overall code rate of 1/3. It is composed of rate 1/2 recursive systematic convolutional (RSC) encoders in which each encoder has two 8-states. They operate in parallel with the data bits, $d_k$, and are interleaved between the two RSC encoders. Various types of code rates can easily be constructed by puncturing various coded bits. In a classic case, parity bits, $x_k^p$ are punctured. In studies [1–4], it is shown that an increment in Hamming distance can be achieved by also puncturing a small number of data bits. Interleaving is a key component of turbo codes. In a turbo encoder, an important parameter is the type of interleaver that is used to increase the maximum likelihood property of coded signals. In general an interleaver is a permutation that maps $N$ input data sequence $\{d_1,d_2,....d_N\}$ into the same sequence of data with a new order of placement. No increment in dimension of the interleaved sequence occurs. If the input data sequence is $d = (d_1,d_2,....d_N)$, then $d_P$ is the permuted data sequence, where $P$ is the interleaving matrix with only one nonzero element equal to one in each column and row. As in all communication schemes, every interleaver has a corresponding deinterleaver that acts on the interleaved data sequence and puts the data back into its original order. The deinterleaving matrix is simply the transposed matrix of interleaving matrix, i.e., $P^T$. So interleaver and deinterleaver accompany each other. Two interleaver types have been commonly studied. They are "random" interleaver and the so-called *S-random* or *spread* interleaver [5–7]. But in ideal conditions, there can never be a exact random interleaver.

## 7.1 Turbo Encoder

Figure 7.1 is a classic diagram of a turbo encoder with the rate of 1/3. Here, $x_k^s$ is the systematic bit, $x_{k1}^p$ and $x_{k2}^p$ are the parity check bits, respectively. The input binary data sequence is expressed by $d_k = (d_1, ..., d_n)$ vector. The input sequence $d_k = (d_1, ..., d_n)$ is passed through a convolutional encoder1 [8,9], and a coded bit stream, $x_{k1}^p$, is formed. The input data sequence is then interleaved. In the block of the interleaver, the input bits are loaded into a matrix and read out in such a way as to spread the positions of the input bits. The bits are often read out in a pseudorandom manner. The critical point is, at the decoder side, similar interleaver/de-interleaver blocks are to be repeated. Thus the interleaver may never be an ideal interleaver with ideal probabilistic properties, so this block is a pseudointerleaver. As in Figure 7.1, an interleaved data sequence is passed through a second convolutional Encoder2, and a second coded bit stream, $x^p_{k_2}$, is generated. Then the outputs of Encoder1 and Encoder2 are punctured and the new set of data is denoted as parity data. Systematic and parity data are modulated, which helps the data passage through the channel with high error performance.

## 7.1.1 RSC Component Codes

Recursive systematic codes (RSC) consist of recursive or feedback blocks, as Encoder1 and Encoder2, and systematic feedforward blocks. In systematic, uncoded data bits appear in the transmitted code bit sequence. A simple RSC encoder is given in Figure 7.2 with rate 1/2, generator polynomial $G = \{g_1, g_2\} = \{7,5\}$ and constraint length $K = 3$. Here $g_1$ is the feedback connectivity and $g_2$ is the output connectivity, in octal formation. The RSC component encoder has generally two output sequences; these are data sequence, $x^s_k = (x^s_1, \ldots, x^s_N)$ and parity sequence $x^p_k = (x^p_1, \ldots, x^p_N)$.

It is necessary to first discuss the structure of error control codes to explain the reason for choosing RSC component codes in a turbo code encoder, rather than the conventional nonsystematic encoder codes (NSC). The minimum number of errors that an encoder can detect or correct is determined by the minimum Hamming distance of the code. Comparison of each possible code word with the all-zeroes code word, shows the minimum Hamming distance of classic turbo codes. Thus, in other words, in turbo codes, minimum code weight (number of "1"s in a code word) chosen from all the possible code words determines also the minimum Hamming distance. The minimum Hamming distance is the main parameter

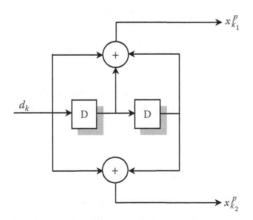

**Figure 7.2   Example of nonsystematic convolutional code.**

in AWGN channels for asymptotic performance of a bit-error rate (BER) at a high signal-to-noise ratio (SNR).

In turbo codes, an error floor occurs because of the very slow fall at high SNRs. It is clear that higher minimum Hamming distance results in a quick rate of fall of BER as SNR takes larger values. It is interesting that at low SNR values, code words with code weights larger than the minimum value are effective in error-performance analysis. This is the relationship between the code weight and the number of code words with that code weight. Thus the overall distance spectrum of the code is important. RSC codes have an infinite-impulse response. If a data sequence consisting of a "1" followed by a series of "0"s enters the RSC encoder, a code sequence will be generated containing both ones and zeroes for as long as the subsequent data bits are zero. It can be concluded that these encoders will tend to generate high weight code sequences, practically correlated with input data. The performance of RSC codes is enhanced by replacing interleaver between encoders {Encoder1 and Encoder2} as in Figure 7.2. An interleaver is a mixing element that produces an alternative sequence with a different ordering of the input sequence. Because the same encoder/decoder blocks are to be used in transmitter and receiver structures in wireless communications—as in every type of communication environment—it is necessary to form the same interleaving/deinterleaving procedure in both sides. Thus turbo codes tend to make use of pseudorandom interleavers instead of real random interleavers, where reproduction is impossible in a short period of time.

In practice, high minimum Hamming distances do not guarantee good asymptotic performance. With high SNR values, pseudorandom mappings lead to a low minimum code weight compared to convolutional codes and leading to a marked error floor for high SNR, because pseudointerleaver can never act as ideal interleaver. The interleaver design is crucial in insuring that a turbo code/interleaver combination has the lowest possible error floor.

At low SNR values, RSC code and interleaving combinations produce codewords with higher code weights. Then, it may be concluded that RSC as a whole becomes significant in determining BER performance at low SNR's. The low code weight sequences associated with turbo codes is known as spectral thinning, and leads to their good BER performance at low SNR [10].

## 7.1.2  Trellis Termination

Trellis termination can be achieved in classic convolutional encoders by inserting $m = K - 1$ additional zero bits after the input sequence and these inserted bits convert to all-zero state. However, this strategy is not possible for the RSC encoder because feedback. The extra termination bits depend

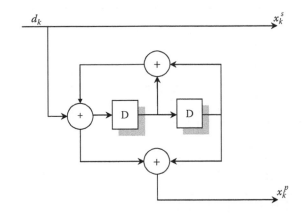

**Figure 7.3   Example of a recursive systematic convolutional (RSC) code.**

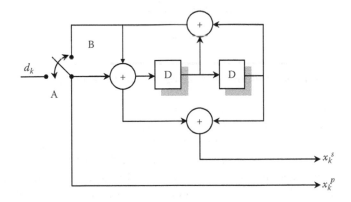

**Figure 7.4   A simple trellis termination modification of a RSC encoder.**

on the state of the RSC encoder and are very difficult to predict. Futhermore, interleaver can also prevent a predicted outcome. As one solution, Figure 7.4 shows a simple modification of Figure 7.3 that does not change the RSC property, while trellis termination occurs.

In the encoding process, the switch is turned to position A and in terminating the trellis, the switch is turned to position B.

## 7.1.3 *Interleaving*

Interleaver is a block in the encoder scheme that scrambles the placement order of the input bits before feeding them into the second encoder as shown in Figure 7.5. In digital communication, BER errors decrease when

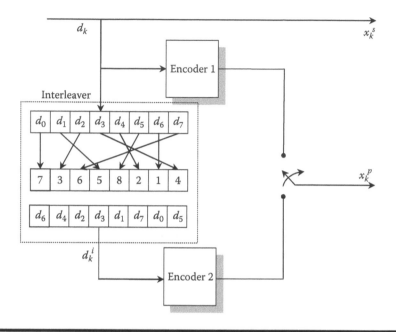

**Figure 7.5  A sample random interleaving in a turbo encoder.**

noisy sequences are received spread far apart. Thus burst error patterns can be broken in the overall code structure. The interleaver has a pseudorandom pattern, known by both receiver and transmitter. As previously stated, interleaver maps input and output positions of the bits randomly. Assume an interleaver of length, the input sequence $i = 0, 1, ..., (N - 1)$ is scrambled according to a pseudorandom number set to form output sequence $\pi(i) = 0, 1, ..., (N - 1)$. In other words, dimensions of the sequences at the entrance and output of the interleaver block are the same. Only the placement of bits is scrambled due to a pseudo-random number set. Scrambling algorithms are easily generated, but the main problem is that they don't guarantee minimum spreading properties and therefore BER performance improvement. A deinterleaver at the receiver reverses the scrambling effect of the interleaver.

In random interleaving turbo encoder, both encoders receive the same input bits with different orders. The original data stream is $X_k^s$, the parity codes are $X_{k1}^p$ and $X_{k2}^p$. After the selection of a puncturer $X_k^p$ data stream represents complementary data information. Thus the overall code rate is 1/3.

### 7.1.4  Convolutional Code Representation

Convolutional code is simply a finite impulse response (FIR) filter with which stability is guaranteed. They are generally considered as finite state machines and modeled by state diagrams and trellis structures. At first, a suitable state diagram is drawn regarding the memory units-shift registers. Then a number of inputs/outputs complete the diagram and define the complexity, because there must be connections between all nodes, representing the values kept in memory units. If the number of memory units is $M$, then there are $2^M$ possible internal states, nodes. The connection lines between $2^M$ nodes are directly correlated with the number of input/output bits among these nodes [11].

### 7.1.4.1 State Diagram Representation

A state diagram is simply composed of nodes and their connections, known as branches. The branches or paths are labeled by $i/o$ where, $i$ represents input information bits and $o$ as output. The input bits are passed through a trellis structure with the current state and it will either change or remain according to the trellis model and input bits. In general the state diagram is terminated by zeros and it is customary to begin convolutional encoding from the all-zero state. As an example, study a state diagram with two shift registers and one input/two outputs (rate $1/2$) structure as shown in Figure 7.6. Assume the input sequence as $i = \{00001011111\}$, then states will be $\{00,00,00,00,10,01,10,11,11,11,11,11\}$ and output encoded sequence $o = \{00,00,00,00,11,10,00,01,10,10,10,10,10\}$.

It is clear that a state diagram makes it easy to understand finite state machine problems with different input/output and state combinations. The trellis structure of the above state transition diagram can easily be constructed by using the similar techniques explained in Section 4.2.1.6.

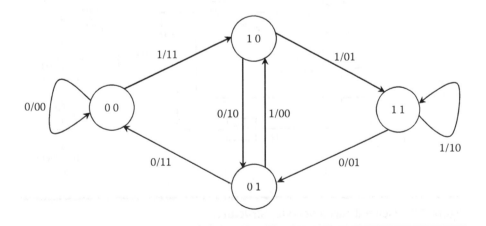

**Figure 7.6   State transitions for input sequence {00001011111}.**

## 7.2 Turbo Decoding

The turbo decoder block is more complex compared to the encoder side. It works in an iterative fashion, as shown in Figure 7.7. The term turbo codes was given because of the feedback from the output as with turbo engines. Iterations show the number of re-processing and it changes with the difficulty of the problem. Of course in communications, difficulty means low SNR, severe fading, MIMO and time varying channels, mobile indoor environments, and so forth. One loop is performed at a time for each iteration [1]. The main disadvantage of turbo decoding is that there are as many as 20 iterations, resulting in delays in the decoding process. As in Figure 7.7, noisy systematic $y_k^s$ and noisy parity $y_k^p$ input sequences are fed into the decoder scheme. After DEMUX, the input sequence is passed to two different soft-input soft-output (SISO) decoder blocks; SISO-DEC1 and SISO-DEC2. SISO-DEC1 and SISO-DEC2 decode sequences formed at the output of ENC1 and ENC2, respectively. All the decoders are based on maximum a posteriori (MAP) algorithm. SISO-DEC1 decodes noisy inputs; received systematic sequence $\{y_k^s\}$ and parity sequence $\{y_{k1}^s\}$. The output of SISO-DEC1 is a sequence of soft estimates of the transmitted data bits $\{d_k\}$ and this output is denoted as extrinsic data EXT1. Since the second SISO-DEC2 inputs are interleaved data, the first decoder output and noisy systematic received values $\{y_k^s\}$ are both interleaved. Thus interleaved systematic received values $\{y_k^s\}$ and the sequence of received

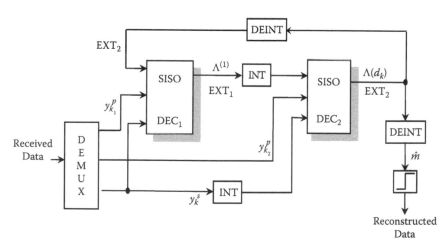

**Figure 7.7 General turbo decoder structure.**

parity values $\{y_{k2}^p\}$ from the second encoder ENC2 and interleaved extrinsic information EXT1 are fed into the second decoder, SISO-DEC2. The output of this decoder is deinterleaved and a hard decision made. If the results of the turbo decoder are not sufficient in the view of BER performance, then SISO-DEC2 output is deinterleaved and fed into the first decoder. Thus, the first iteration is processed. The iterations continue untill significant performance is obtained in the decoding stage. For example, Berrou and Glavieux [1] achieved BER = $10^{-5}$ at SNR within 0.7 dB of the Shannon limit after 18 iterations.

## 7.2.1 MAP Algorithm

As explained in Section 7.2, the decoders SISO-DEC1 and SISO-DEC2 use the MAP algorithm to estimate information data. The MAP algorithm estimates *a posteriori* probability (APP) of each state transition, information bit, produced by a Markov process, given the noisy received signal **y**. After calculation of APPs for all possible values of the desired quantity, a hard decision is made by taking the quantity with highest probability. The MAP algorithm calculates the APPs of message bits $P[m_i = 1|y]$ and $P[m_i = 0|y]$, which are then put into LLR (log likelihood ratio) form according to

$$\Lambda_i = \ln \frac{P_i[m_i = 1|y]}{P_i[m_i = 0|y]}.$$

Here, $m_i$ is the message bit associated with the state transitions and $P[.]$ is the conditional probability.

In turbo coding, the MAP algorithm makes hard decisions on only information bits at the last decoder iteration. The MAP algorithm first finds the probability $P[s_i \rightarrow s_{i+1}|y]$ of each valid state transition given the noisy channel observation **y** as

$$P[s_i \rightarrow s_{i+1}|\mathbf{y}] = \frac{P[s_i \rightarrow s_{i+1}, \mathbf{y}]}{P[\mathbf{y}]} \tag{7.1}$$

where $s_i$ represents the state of the trellis at the time $i$. The conditional probability $P[.]$ is calculated for state transition $s_i$ to $s_{i+1}$. The numerator can be rewritten as

$$P\left[ s_i \rightarrow s_{i+1}, y \right] = \alpha(s_i)\gamma(s_i \rightarrow s_{i+1})\beta(s_{i+1}),$$

$$\alpha(s_i) = P\left[ s_i, (y_0, \ldots\ldots y_{i-1}) \right]$$

$$\gamma(s_i \rightarrow s_{i+1}) = P\left[ s_{i+1}, y_i \,\middle|\, s_i \right] \qquad (7.2)$$

$$\beta(s_{i+1}) = P\left[ (y_{i+1}, \ldots\ldots, y_{L-1}) \,\middle|\, s_{i+1} \right].$$

The term $\gamma(s_i \rightarrow s_{i+1})$ is the branch metric associated with the transition $s_i \rightarrow s_{i+1}$, which can be expressed as

$$\gamma(s_i \rightarrow s_{i+1}) = P\left[ s_{i+1} \,\middle|\, s_i \right] P\left[ y_i \,\middle|\, s_i \rightarrow s_{i+1} \right] = P\left[ m_i \right] P\left[ y_i \,\middle|\, x_i \right] \quad (7.3)$$

where $m_i$ and $x_i$ are the message and output associated with the state transition $s_i \rightarrow s_{i+1}$. When there is no connection between any two consecutive $s_i$ and $s_{i+1}$ in the trellis diagram, then the above probability is taken as zero [11]. The first term on the right-hand side of the above equation is obtained using the *a priori* information $z_i$ and

$$P\left[ m_i \right] = \begin{cases} \dfrac{e^{z_i}}{1+e^{z_i}} & \text{for } m_i = 1 \\[3mm] \dfrac{1}{1+e^{z_i}} & \text{for } m_i = 1 \end{cases}, \qquad (7.4)$$

$$\ln P\left[ m_i \right] = z_i m_i - \ln(1 + e^{z_i}).$$

The second term is a function of the modulation and channel model. For flat-fading channel,

$$P\left[ y_i \,\middle|\, x_i \right] = \frac{1}{\sqrt{\pi N_0 / E_s}} \exp\left\{ \frac{-E_s}{N_0} \sum_{q=0}^{n-1} \left[ y_i^q - a_i^{(q)}(2x_i^{(q)} - 1) \right]^2 \right\}. \quad (7.5)$$

The probability $\alpha(s_i)$ can be found according to the forward recursion (Figure 7.8).

$$\alpha(s_i) = \sum_{s_{i-1}} \alpha(s_{i-1})\gamma(s_{i-1} \rightarrow s_i) \qquad (7.6)$$

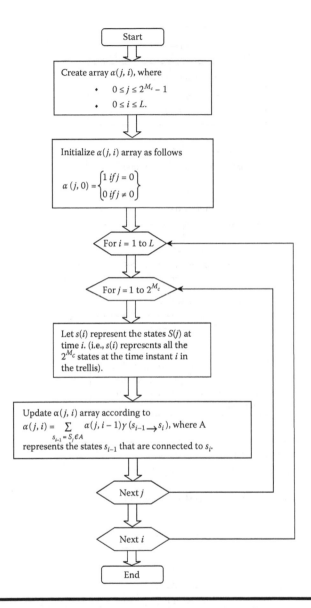

**Figure 7.8   Diagram of forward recursion for MAP algorithm.**

Likewise, $\beta(s_i)$ can be found according to the backward recursion (Figure 7.9).

$$\beta(s_i) = \sum_{s_{i-1}} \beta(s_{i-1})\gamma(s_i \rightarrow s_{i+1}) \tag{7.7}$$

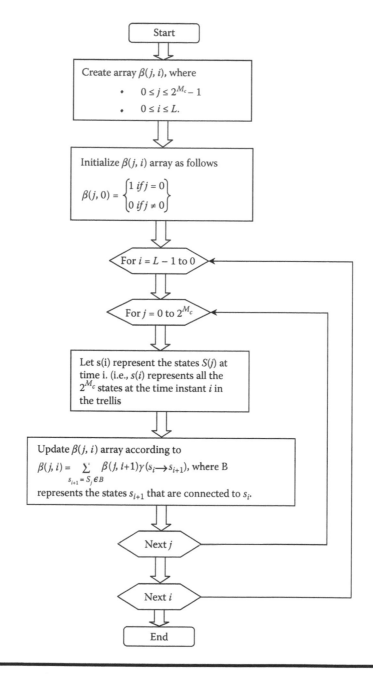

**Figure 7.9  Diagram of backward recursion for MAP algorithm.**

Once the *a posteriori* probability of each state transition $P\left[s_i \rightarrow s_{i+1} \mid y\right]$ is found, the message bit probabilities can be found according to

$$P\left[m_i = 1 \mid y\right] = \sum_{S_1} P\left[s_i \rightarrow s_{i+1} \mid y\right] \tag{7.8}$$

$$P\left[m_i = 0 \mid y\right] = \sum_{S_0} P\left[s_i \rightarrow s_{i+1} \mid y\right]$$

$S_1 = \left\{s_i \rightarrow s_{i+1} : m_i = 1\right\}$ is the set of all state transitions associated with a message bit of 1, and $S_0 = \left\{s_i \rightarrow s_{i+1} : m_i = 0\right\}$ is the set of all state transitions associated with a message bit of 0. The log-likelihood ratio can be written as

$$\Lambda_i = \ln \frac{\displaystyle\sum_{S_1} \alpha(s_i)\gamma(s_i \rightarrow s_{i+1})\beta(s_{i+1})}{\displaystyle\sum_{S_0} \alpha(s_i)\gamma(s_i \rightarrow s_{i+1})\beta(s_{i+1})} \tag{7.9}$$

To summarize, the MAP algorithm can be divided into four major steps:

1. Forward recursion: $\alpha$'s are calculated in this step.
2. Backward recursion: $\beta$'s are calculated in this step.
3. Log-likelihood ratios: The log-likelihood ratios of all the decoded message bits from $i = 0$ to $L - 1$ are calculated in this step as shown in Equation (7.9).
4. Computation of LLR's
   For all $i = (0,1,...L - 1)$, LLR can be found using the following formula (7.9):

$$\Lambda_i = \ln \frac{\left\{\displaystyle\sum_{S_1} \alpha(j,i)\gamma(s_i \rightarrow s_{i+1})\beta(j',i+1)\right\}}{\left\{\displaystyle\sum_{S_0} \alpha(j,i)\gamma(s_i \rightarrow s_{i+1})\beta(j',i+1)\right\}} \tag{7.10}$$

$S_0$ represents all the state transitions for which the input message bit is "0" and $S_1$ represents the set of all state transitions for which the input message bit is "1".

## 7.2.2 Max-Log-MAP and Log-MAP Algorithms

*A posteriori* probability of each message bit can be calculated precisely using the well-known MAP algorithm, but it is computationally intensive, requiring $6 \times 2^{Mc}$ multiplications per estimated bit and an equal number of additions. It is also sensitive to round-off errors. Choosing the entire algorithm in the log-domain study will minimize these two key obstacles.

In logarithm space all multiplications are replaced with additions. Thus both in time and complexity domain-severe simplicity can be achieved. Compare the performance of log-domain for Jacobian logarithm

$$\ln(e^x + e^y) = \max(x, y) + \ln(1 + \exp\{-|y - x|\})$$
$$= \max(x, y) + f_c(|y - x|). \tag{7.11}$$

In log-domain, addition turns out to be a simple maximization operation followed by a correction function $f_c\{.\}$ and, of course, no exponential values and multiplication operations. Note that in the case of a great absolute difference between $y$ and $x$, the correction function tends to zero. As a result of this special case, Equation (7.11) becomes

$$\ln(e^x + e^y) \approx \max(x, y) \tag{7.12}$$

Both max-log-MAP and log-MAP algorithms [11,12] of log-domain representations replace multiplications with simple additions. The difference between these algorithms is their computation techniques of addition operations. The max-log-MAP approximates addition in the log-domain as a maximization operation as given in Equation (7.12). The log-MAP algorithm is an elaboration process over the max-log-MAP algorithm, and computes addition in the log domain as a maximization followed by a correction term, as shown in Equation (7.11). The correction function can be tabulated as a one-dimensional look-up table. In [11,13], it is shown that eight entries in the look-up table are enough for general problems. The max-log-MAP algorithm is a generalization form of the soft output Viterbi algorithm (SOVA) as stated in [11,14].

Let $\tilde{\gamma}(s_i \to s_{i+1})$ map to the natural logarithm of $\gamma(s_i \to s_{i+1})$

$$\tilde{\gamma}(s_i \to s_{i+1}) = \ln \gamma(s_i \to s_{i+1})$$
$$= \ln P[y_i|x_i] + \ln P[m_i]. \tag{7.13}$$

In a flat-fading channel with BPSK modulation type, (7.13) can be rewritten as

$$\tilde{\gamma}(s_i \to s_{i+1}) = \lambda(s_i \to s_{i+1}) = \ln P[m_s] - \frac{1}{2}\ln(\pi N_0 / E_s)$$
$$+ \left\{ \frac{-E_s}{N_0} \sum_{q=0}^{n-1} [y_i^q - a_i^{(q)}(2x_i^{(q)} - 1)]^2 \right\} \tag{7.14}$$

Define

$$\tilde{\alpha}(s_i) = \max_{s_{i-1} \in A}{}' [\tilde{\alpha}(s_{i-1}) + \tilde{\gamma}(s_{i-1} \to s_i)]$$
$$= \ln \alpha(s_i) = \ln \left\{ \sum_{s_{i-1}} \propto \text{p}[\tilde{\alpha}(s_{i-1}) + \tilde{\gamma}(s_{i-1} \to s_i)] \right\} \tag{7.15}$$

A is the set of previous states $\{s_{i-1}\}$ that are connected to consecutive state $\{s_i\}$. Max$'(x,y)$ can be expressed as max$'(x,y)$ = max$(x,y)$ for max-log-MAP algorithm and max$'(x,y)$ = max$(x,y)$ + fc$(|y-x|)$ for log-MAP algorithm. Thus max$'(x,y)$ has a different expression for these log-MAP algorithms. In max-log-MAP, the iterative computation of $\tilde{\alpha}(s_i)$ is similar to the Viterbi algorithm. In log-MAP algorithm, $\tilde{\alpha}(s_i)$ is computed using a modified version of the Viterbi algorithm, where only a correction term $f_c\{.\}$ is added. After the logarithm of forward term $\alpha(s_i)$ as in Equation (7.15), we now focus on logarithm of backward term $\beta(s_i)$ as

$$\tilde{\beta}(s_i) = \max_{s_{i+1} \in B}{}' [\tilde{\beta}(s_{i+1}) + \tilde{\gamma}(s_i \to s_{i+1})] = \ln \beta(s_i)$$
$$= \ln \left\{ \sum_{s_{i+1}} \exp[\tilde{\beta}(s_{i+1}) + \tilde{\gamma}(s_i \to s_{i+1})] \right\} \tag{7.16}$$

where $\beta$ is the set of states $\{s_{i+1}\}$ that are connected to state $\{s_i\}$. Equation (7.16) can be calculated using the Viterbi algorithm or the generalized Viterbi algorithm [11]. After $\tilde{\alpha}(s_i)$ and $\tilde{\beta}(s_i)$ have been estimated for all possible states in the trellis as in Figures 7.10 and 7.11, the log likelihood ratio (LLR) can be written as follows;

$$
\begin{aligned}
\Lambda_i = \ln\left\{\sum_{S_1} \exp\left[\tilde{\alpha}(s_i) + \tilde{\gamma}(s_i \rightarrow s_{i+1}) + \tilde{\beta}(s_{i+1})\right]\right\} \\
- \ln\left\{\sum_{S_0} \exp\left[\tilde{\alpha}(s_i) + \tilde{\gamma}(s_i \rightarrow s_{i+1}) + \tilde{\beta}(s_{i+1})\right]\right\} \\
= \max_{S_1}' \left[\tilde{\alpha}(s_i) + \tilde{\gamma}(s_i \rightarrow s_{i+1}) + \tilde{\beta}(s_{i+1})\right] \\
- \max_{S_0}' \left[\tilde{\alpha}(s_i) + \tilde{\gamma}(s_i \rightarrow s_{i+1}) + \tilde{\beta}(s_{i+1})\right]
\end{aligned}
\tag{7.17}
$$

The max-log-MAP and log-MAP algorithms first compute the branch metric $\tilde{\gamma}$ according to (7.14), then forward recursion $\alpha$, and backward recursion $\beta$, are computed according to the diagrams given in Figures 7.10 and 7.11; finally LLRs are computed as below.

### Computing the LLRs

For $i = (0,1,..., L - 1)$, find out the LLR using the following formula:

$$
\begin{aligned}
\Lambda_i = \max_{S_1}^* \sum_{S_1} \tilde{\alpha}(j,i) + \tilde{\gamma}(s_i \rightarrow s_{i+1}) + \tilde{\beta}(j,i+1) \\
- \max_{S_0}^* \sum_{S_0} \tilde{\alpha}(j,i) + \tilde{\gamma}(s_i \rightarrow s_{i+1}) + \tilde{\beta}(j,i+1)
\end{aligned}
\tag{7.18}
$$

$S_1 = (s_i = S_j) \rightarrow (s_{i+1} = S_j):m_i = 1$ defines a set of all state transitions for which message bit of "1" and in the same manner $S_0 = (s_i = S_k) \rightarrow (s_{i+1} = S_k):m_i = 0$ is the set of all the state transitions associated with a message bit is "0" [11].

### 7.2.3 Performance Analysis

In AWGN, bit error performance of a $(n,k)$ block codes can be union bounded as

$$
P_b \leq \sum_{i=1}^{2^k-1} \frac{w(\mathbf{m}_i)}{k} Q\left(\sqrt{\frac{w(\mathbf{x}_i)2rE_b}{N_o}}\right)
\tag{7.19}
$$

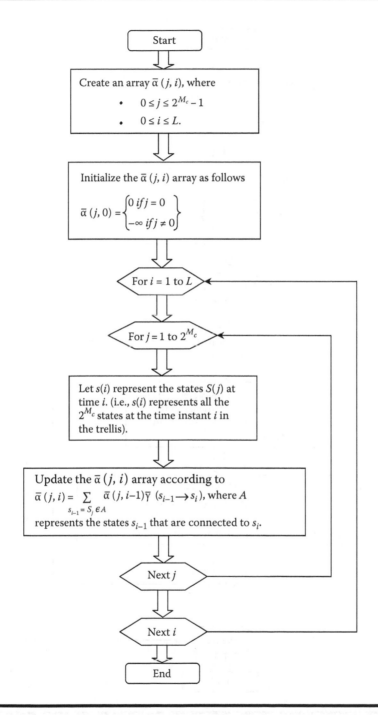

**Figure 7.10** **Diagram of forward recursion for max-log-MAP algorithm.**

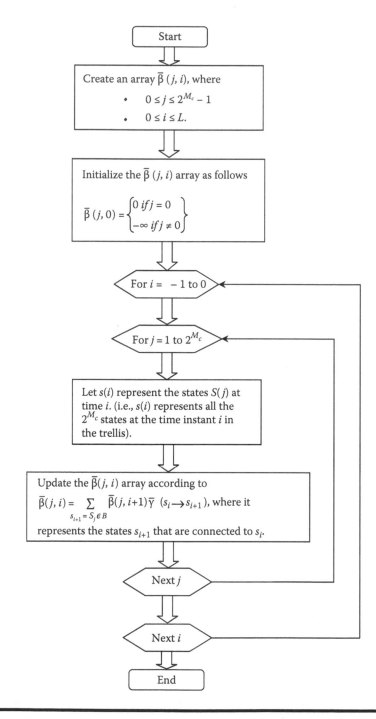

**Figure 7.11  Diagram of backward recursion for max-log-MAP algorithm.**

where the $Q$-function is defined by

$$Q(z) = \frac{1}{\sqrt{2\pi}} \int_z^{\infty} e^{\frac{-x^2}{2}} dx$$

and $E_b$ is the energy per bit, and $N_o$ is the one-sided noise spectral density. We can compute Equation (7.19), if we know the weights of all $2^k - 1$ nonzero code words and their corresponding messages. The complexity increases as $k$ gets larger. In the case of gathering terms producing the same code word weights, we can rewrite Equation (7.19) as

$$P_b \leq \sum_{d=d_{\min}}^{n} \frac{\overline{w}_d N_d}{k} Q\left(\sqrt{\frac{d\, 2rE_b}{N_o}}\right) \tag{7.20}$$

where $N_d$ is the number of code words of weight $d$, $\overline{w}_d$ is the average weight of the $N_d$ messages that produce weight $d$ code words and $d_{\min}$ asymptotic minimum distance.

## 7.3 Error Performance of Turbo Codes

The main emphasis of the channel model is the performance of turbo coding in a fading environment modeled by Rician probability density function. Thus, the following reflect the degradations from fading on the amplitude of the received signal.

### 7.3.1 Error Performance of Turbo Codes over Fading Channel

At $i$th coding step, if the output of the transmitter $x(i)$, the channel output $y(i)$, is

$$y(i) = b(i) \cdot x(i) + n(i) \tag{7.21}$$

$n(i)$ is Gaussian noise where the noise variance is

$$\sigma^2 = \frac{N_O}{2E_S}$$

and $b(i)$ is complex channel parameter, defined as $b(i) = |\rho(i)| e^{j\theta(i)}$. The fading parameter $\rho(i)$ can be modeled as Rician probability density function (pdf) and Equation (5.4) can be rewritten as,

$$P(\rho) = 2\rho(1 + K) \ e^{(-\rho^2(1+K) - K)} I_o\left[2\rho\sqrt{K(1+K)}\right] \quad \rho \geq 0 \quad (7.22)$$

where $I_o$ is the modified Bessel function of the first kind, order zero and $K$ is a fading parameter. Rician turns into Rayleigh if parameter $K$ is chosen as 0. The phase term of the complex fading parameter $\theta(i)$ can be modeled as Tikhonov distributions as in [15,16],

$$p(\theta) = \frac{e^{\alpha\cos(\theta)}}{2\pi I_o(\alpha)} \qquad |\theta| \quad \pi \qquad (7.23)$$

where $\alpha$ is the effective signal-to-noise ratio in the carrier tracking loop. The BER versus SNR as $E_b/N_0$ curves of turbo-coded signals over various channel parameters are given in Equations 7.12–7.33. In ideal channel state information (CSI) case, channel state parameters are assumed to be known. In all curves ideal CSI are considered. In Figures 7.12–7.16, phase term $\theta(i)$ is not effective, therefore, it is assumed to be zero. The BER improves while the iteration number increases. Iteration means the number of feedback connection (EXT2) as in Figure 7.8. In Figure 7.12, BER performance is drawn for AWGN channel, which corresponds to ($K = \infty$). In this chapter, all figures are evaluated for a frame size of $N = 400$, in which the data set is grouped in 400 bits. It is clear that as $N$ gets larger, BER will be improved. The numerical results clearly demonstrate that the error performance degradation is due to both amplitude and phase noise process. It is clear that phase jitter distortion is effective both in severe fading ($K = 0$ dB) and also Rician fading.

In Figure 7.16, the iteration number is kept constant as the 5th iteration and error performances are compared for various fading parameter values $K = 0,10,20,\infty$ (dB).

In Figure 7.16, at 1 dB SNR, they concur, but after 2 dB, differentiation of channel parameters becomes effective. As an example, at $P_b = 10^{-3}$, about 4dB gain occurs between Rayleigh ($K = 0$) and AWGN ($K = \infty$).

In the case of an imperfect phase reference defined in Equation (7.23) as the Tikhonov distribution, turbo-coded signals are simulated for BER versus SNR, $\alpha$ values. In these curves, SNR and $\alpha$ values act simultaneously. The x-axis shows SNR and $\alpha$ both in dB and y-axis shows BER in Figures 7.17–7.21.

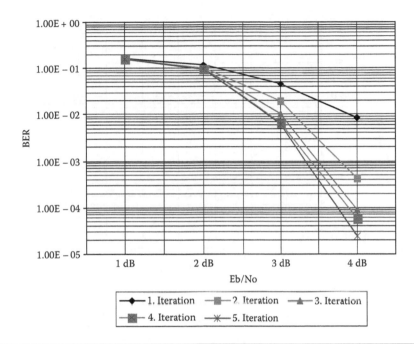

**Figure 7.12  Turbo coding in AWGN channel ($K = \infty$).**

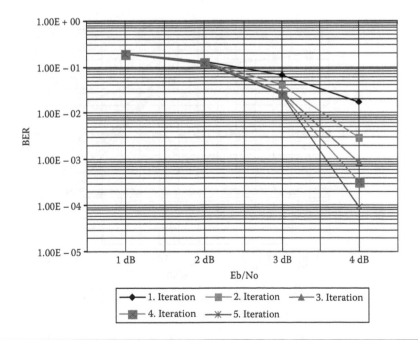

**Figure 7.13  Turbo coding in Rician channel ($K$ = 20 dB).**

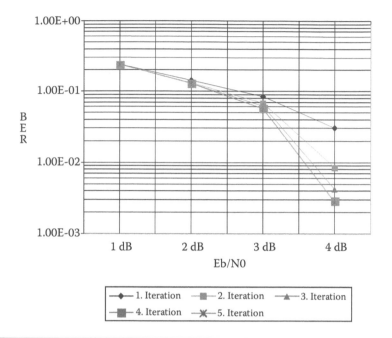

**Figure 7.14    Turbo coding in Rician channel (*K* = 10 dB).**

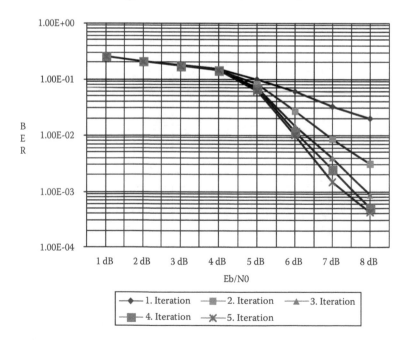

**Figure 7.15    Turbo coding in Rayleigh channel (*K* = 0 dB).**

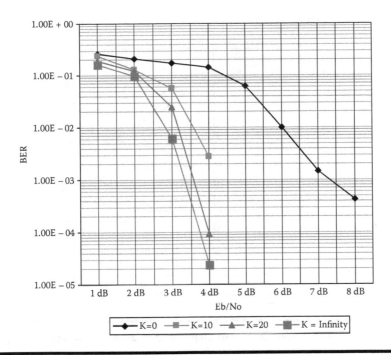

**Figure 7.16    Turbo coding in fading channel for *K* = 0, 10, 20, ∞ (dB), for 5th iteration.**

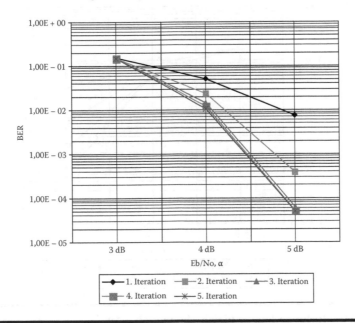

**Figure 7.17    Turbo coding in AWGN channel (*K* = ∞ dB) with imperfect phase reference.**

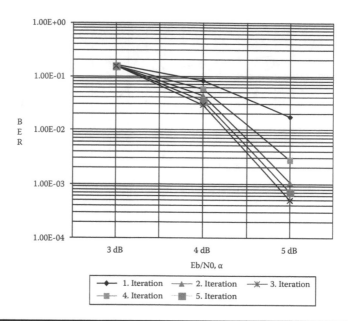

**Figure 7.18** Turbo coding in Rician channel ($K$ = 20 dB) with imperfect phase reference.

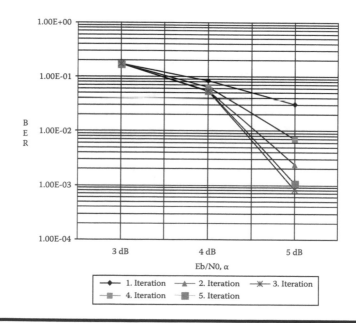

**Figure 7.19** Turbo coding in Rician channel ($K$ = 10 dB) with imperfect phase reference.

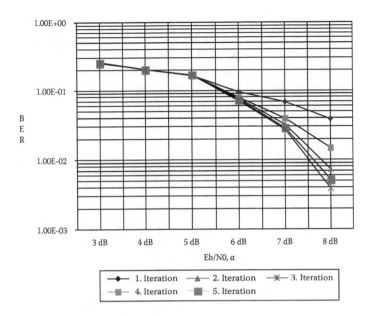

**Figure 7.20    Turbo coding in Rayleigh channel (*K* = 0 dB) with imperfect phase reference.**

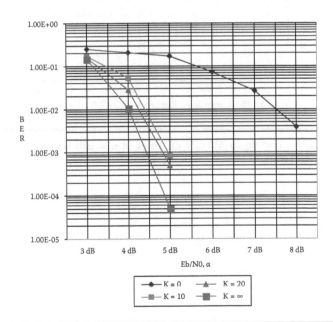

**Figure 7.21    The performance of turbo coding with imperfect phase reference for K = 0, 10, 20, ∞ (dB) for 5th iteration.**

As shown in Figure 7.21, for a constant iteration number as *K* increases, error performance improves simultaneously for SNR and α parameters.

In the following graphics, BER is simulated for a different α phase jitter parameter in various fading environments (see Figures 7.22–7.33). A point of interest is the effect of the jitter parameter on bit error performance. The jitter effect becomes more effective at higher SNR values for all *K* values.

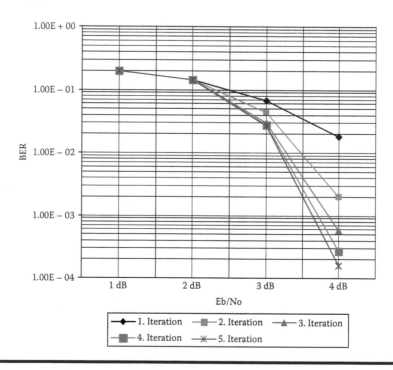

**Figure 7.22  Turbo coding in AWGN channel (*K* = ∞ dB) with imperfect phase reference for α = 10 dB.**

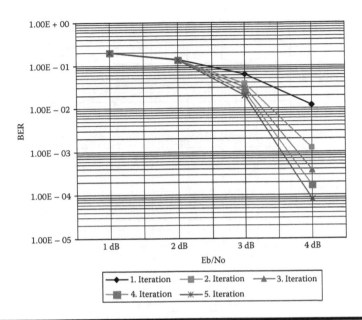

**Figure 7.23** **Turbo coding in AWGN channel ($K = \infty$ dB) with imperfect phase reference for $\alpha = 20$ dB.**

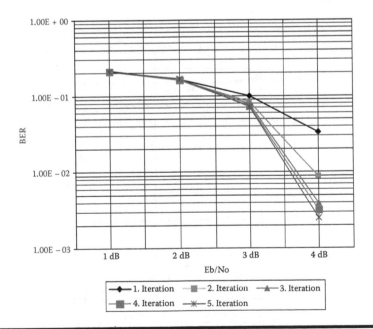

**Figure 7.24** **Turbo coding in Rician channel ($K = 20$ dB) with imperfect phase reference for $\alpha = 10$ dB.**

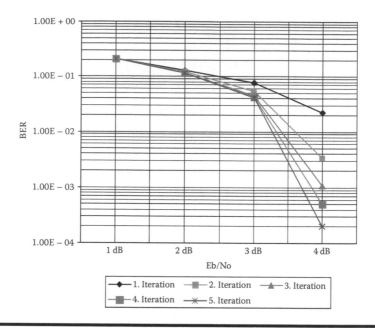

**Figure 7.25  Turbo coding in Rician channel ($K$ = 20 dB) with imperfect phase reference for $\alpha$ = 20 dB.**

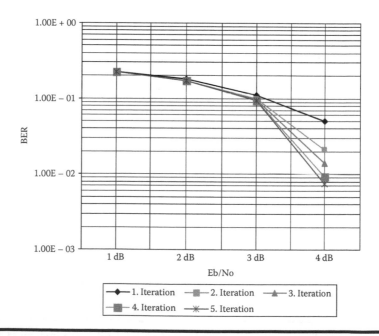

**Figure 7.26  Turbo coding in Rician channel ($K$ = 10 dB) with imperfect phase reference for $\alpha$ = 10 dB.**

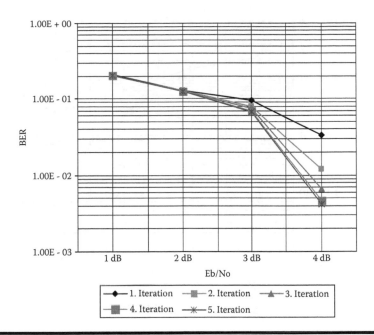

**Figure 7.27 Turbo coding in Rician channel (*K* = 10 dB) with imperfect phase reference for α = 20 dB.**

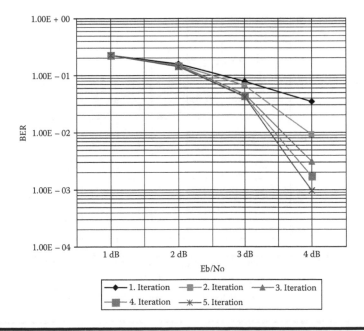

**Figure 7.28 Turbo coding in Rayleigh channel with imperfect phase reference for α = 10 dB.**

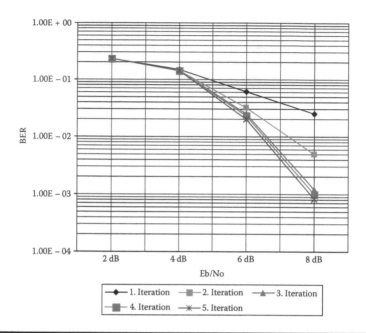

**Figure 7.29  Turbo coding in Rayleigh channel with imperfect phase reference for α = 20 dB.**

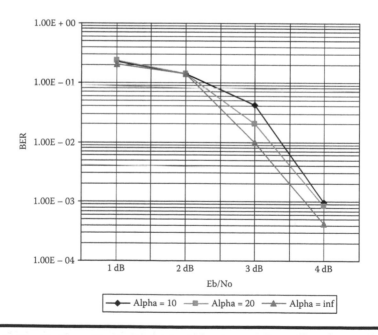

**Figure 7.30  Turbo coding in Rayleigh channel (*K* = 0 dB) with various imperfect phase references.**

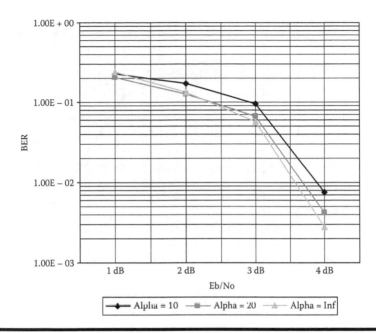

**Figure 7.31   Turbo coding in Rician channel (*K* = 10 dB) with various imperfect phase references.**

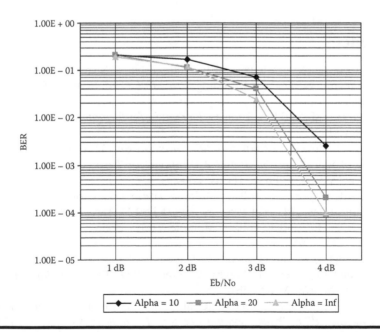

**Figure 7.32   Turbo coding in Rician channel (*K* = 20 dB) with various imperfect phase references.**

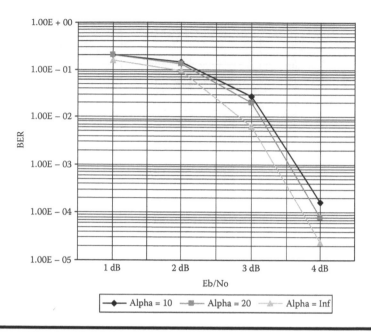

**Figure 7.33** **Turbo coding in AWGN channel (*K* = 0 dB) with imperfect phase reference.**

# References

[1] Berrou, C., A. Glavieux, and P. Thitimajshima. Near Shannon Limit Error Correcting Coding and Decoding. *Proc. Intl. Conf. Comm.*, 1064–1070, May 1993.

[2] Cover, T. M., and J. A. Thomas. *Elements of Information Theory.* New York: Wiley, 1991.

[3] Elias, P. Error-Free Coding. *IEEE Trans. Inform. Theory*:29–37, Sept. 1954.

[4] Buyukatak, K. *Evaluation of 2-D Images for Turbo Codes.* Istanbul Technical University, Ph.D. Thesis, 2004.

[5] Divsalar, D., and F. Pollara. Multiple Turbo Codes for Deep-Space Communications. *JPL, TDA Progress Report*:42–121, May 15, 1995.

[6] Divsalar, D., and F. Pollara. Multiple Turbo Codes. *MILCOM'95*:279–285, Nov. 6-8, 1995.

[7] Benedetto, S., and G. Montorsi. Unveiling Turbo Codes: Some Results on Parallel Concatenated Coding Schemes. *IEEE Trans. on Inform. Theory* 42(2):409–428, March 1996.

[8] Viterbi, A. J. Convolutional Codes and Their Performance in Communication Systems. *IEEE Trans. Communication Technology* COM-19(15):751–772, Oct. 1971.

[9] Sklar, B. *Digital Communications: Fundamentals and Applications.* Upper Saddle River, NJ: Prentice Hall International, 1988.

[10] Perez, L, J. Seghers, and D. Costello. A Distance Spectrum Interpretation of Turbo Codes. *IEEE Transactions on Information Theory*. 42(6):1698–1709, Nov. 1996.

[11] Valenti, M. C. Iterative Detection and Decoding for Wireless Communications. A Proposal for Current and Future Work toward Doctor of Philosophy Degree, Sept. 1998.

[12] Robertson, P., P.Hoener, and E. Villebrun. Optimal and Sub-optimal Maximum a Posteriori Algorithms Suitable for Turbo Decodings. *European Trans. on Telecommun.* 8:119–125, Mar./Apr. 1977.

[13] Fossorier, M. P. C., F. Burkert, S. Lin, and J. Hagenauer. On the Equivalence between SOVA and Max-Log-MAP Decodings. *IEEE Commun. Letters* 2:137–139, May 1998.

[14] Fossorier, M. P. C., F. Burkert, S. Lin and J. Hagenauer. On the Equivalence between SOVA and Max-Log-MAP Decodings. *IEEE Commun. Letters* 2:137–139, May 1998.

[15] U•an, O. N., O.Osman, and A.Gümü. Performance of Turbo Coded Signals over Partial Response Fading Channels with Imperfect Phase Reference. *Istanbul University-Journal of Electrical & Electronics Engineering (IU-JEEE)* 1(2):149–168, 2001.

[16] U•an, O. N., and O. Osman. *Communication Theory and Engineering Applications* (in Turkish). Istanbul: Nobel Press, 2006.

# Chapter 8

# Multilevel Turbo Codes

The challenge to find practical decoders for large codes was not encountered until turbo codes, introduced by Berrou et al. in 1993 [1]. The performance of these new codes is close to the Shannon limit with relatively simple component codes and large interleavers. Turbo codes are a new class of error correction codes introduced along with a practical decoding algorithm. Turbo codes are important because they enable reliable communications with power efficiencies close to the theoretical limit predicted by Claude Shannon. They are the most efficient codes for low-power applications such as deep-space and satellite communications, as well as for interference limited applications such as third generation cellular and personal communication services. Because turbo codes use convolutional codes as their constituent codes, a natural extension of the turbo concept, which improves bandwidth efficiency, is its application to systems using TCM. The main principle of turbo codes, given in [1], is conveyed to TCM and this new scheme is denoted as turbo trellis-coded modulation (TTCM). Hence, the important properties and advantages of both structures are retained. Just as binary turbo codes, TTCM uses a parallel concatenation of two binary recursive convolutional encoders, two recursive TCM encoders are concatenated, and interleaving and puncturing are adapted as in [2,3].

In trellis-based structures, to improve the bit error probability, many scientists not only study the channel parameters as in [4,5] but as in [6–9] they have also used multilevel coding as an important band and power efficient technique because it provides a significant amount of encoding gain and low coding complexity. Currently, there are also many attempts

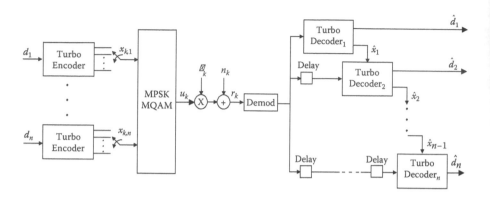

**Figure 8.1  ML-TC structure block diagram. (Copyright permission from John Wiley & Sons [10].)**

to improve the performance of turbo-based systems. For the same purpose, a new type of turbo code known as multilevel turbo codes (ML-TC) have been developed by researches of [9–12].

ML-TC maps the outputs of more than one encoder into a related symbol according to the new set partitioning technique, and each encoder is defined as a level. Since the number of encoders and decoders are the same for each level of the multilevel encoder there is a corresponding decoder, defined as a stage. Furthermore, except in the first decoding stage, the information bits, which are estimated from the previous decoding stages, are used for the next stages as in Figure 8.1.

Multilevel turbo coding maps the outputs of more than one encoder into a related symbol according to the new set partitioning technique and each encoder is defined as a level. Because the number of encoders and decoders are the same for each level of the multilevel encoder there is a corresponding decoder defined as a stage. Furthermore, except for the first decoding stage, the information bits, estimated from the previous decoding stages, are used for the next stages.

## 8.1  Multilevel Turbo Encoder

The basic idea of multilevel coding is to partition a signal set into several levels and to encode each level separately by a proper component of the encoder. First partitioning levels are very effective in the performance of ML-TC schemes. An ML-TC system contains more than one turbo encoder/decoder blocks in its structure. The parallel input data sequences are encoded by our multilevel scheme and mapped to any modulation type such as MPSK, MQAM and so forth. Then, these modulated signals

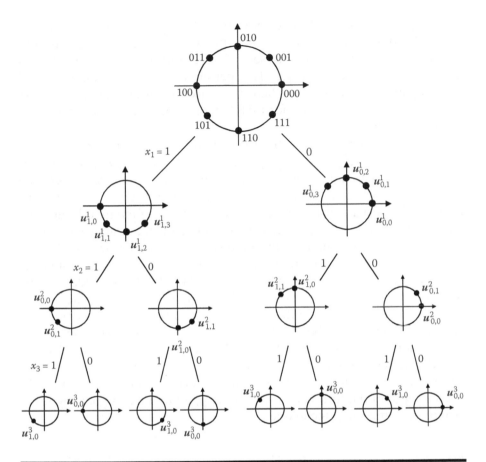

**Figure 8.2  Set partitioning of three-level 8-PSK multi level codes (Copyright permission from John Wiley & Sons [10].)**

are passed through narrow-band fading channels. A multilevel turbo encoder and decoder consist of many parallel turbo encoder/decoder levels as in Figure 8.1. At the receiver, there is one binary turbo encoder at every level of multilevel turbo encoder. To begin, an input bit stream is converted from serial to parallel. Each turbo encoder has only one input and encoding is processed simultaneously. The outputs of these encoders can be punctured and then mapped to any modulated scheme using the group-set partitioning technique as in Figure 8.2. Here, if the $d_k$ is the input of multilevel encoder at time $k$, the encoder output $x_k$ is equal to

$$x_k = f(d_k) \tag{8.1}$$

where $f(.)$ corresponds to the encoder structure. In group partitioning, for the first-level encoder, $x_{k,1}$ is the output bit of the first-level turbo encoder where the signal set is divided into two subsets. If $x_{k,1} = 0$ then the first subset is chosen, and if $x_{k,1} = 1$ then the second subset is chosen. The $x_{k,2}$ bit is the output bit of the second level turbo encoder and divides the subsets into two as before. This partitioning process continues until the last partitioning level is reached. To clarify the partitioning in multilevel turbo coding, the signal set for 8-PSK with three-level turbo encoder is chosen as in Figure 8.2.

Here, if the output bit of the first-level turbo encoder is $x_{k,1} = 0$, then $u_0^1$ set $\{u_{0,0}^1, u_{0,1}^1, u_{0,2}^1, u_{0,3}^1\}$, if $x_{k,1} = 1$, then $u_1^1$ set $\{u_{1,0}^1, u_{1,1}^1, u_{1,2}^1, u_{1,3}^1\}$ is chosen. The output bits of the second-level turbo encoder determine whether $u_1^2$ $\{u_{1,0}^2, u_{1,1}^2\}$ or $u_0^2$ $\{u_{0,0}^2, u_{0,1}^2\}$ subsets are chosen. In the last level, $u_0^3$ $\{u_{0,0}^3\}$ signal is transmitted if the output bit of the third-level turbo encoder is 0, and $u_1^3$ $\{u_{1,0}^3\}$ signal is transmitted if the output bit is 1. In Figure 8.3, for example, a multilevel-turbo coding system for 4-PSK two-level turbo codes is shown. In each level, a 1/3 recursive systematic convolutional (RSC) encoder with memory size $M = 2$ is considered. The bit sequence is converted from serial to parallel. Then, for each level, they are encoded by the turbo encoders. At turbo encoder outputs, the encoded bit streams to be mapped to M-PSK signals are determined after a puncturing procedure. The first bit is taken from the first-level turbo encoder output, the second bit is taken from second-level encoder output, and the other bits are obtained similarly. Following this process, the bits at the output of the turbo encoders are mapped to 4-PSK signals by using group set partitioning technique.

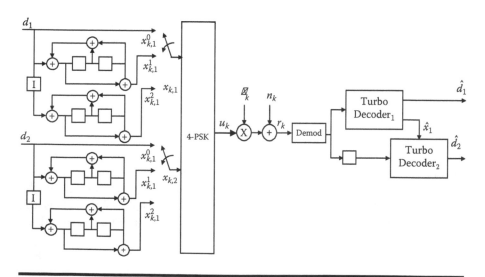

**Figure 8.3  4-PSK two level-turbo coding structure. (Copyright permission from John Wiley & Sons [10].)**

As an example, the remainder $w_k$ of the encoder can be found using feedback polynomial $g^{(0)}$ and feedforward polynomial $g^{(1)}$. The feedback variable is

$$w_k = d_k + \sum_{j=1}^{K} w_{k-j} g_j^{(0)} \tag{8.2}$$

RSC encoders outputs $x_k^{1,2}$, which are called parity data, are

$$x_k^{1,2} = \sum_{j=0}^{K} w_{k-j} g_j^{(1)} \tag{8.3}$$

In our case, RSC encoder has feedback polynomial $g^{(0)} = 7$ and feedforward polynomial $g^{(1)} = 5$, and it has a generator matrix

$$G(D) = \left[ 1 \quad \frac{1 + D + D^2}{1 + D^2} \right] \tag{8.4}$$

where $D$ is memory unit.

## 8.2 Multilevel Turbo Decoder

On the receiver side, the input sequence of the first level is estimated from the first turbo decoder. Subsequently, the other input sequences are computed by using the estimated input bit streams of previous levels. The problem of decoding turbo codes involves the joint estimation of two Markov processes, one for each constituent code. While in theory, it is possible to model a turbo code as a single Markov process, doing so is extremely complex and does not lend itself to computationally tractable decoding algorithms. In turbo decoding, there are two Markov processes that are defined by the same set of data, hence, the estimation can be refined by sharing the information between the two decoders in an iterative fashion. More specifically, the output of one decoder can be used as *a priori* information by the other decoder. If the outputs of the individual decoders are in the form of hard-bit decisions, then there is little advantage in sharing information. However, if soft-bit decisions are produced by the individual decoders, considerable performance gains can be achieved by executing multiple iterations of decoding. The received signal can be shown as

$$r_k = \rho_k u_k + n_k \tag{8.5}$$

where $r_k$ is a noisy received signal, $u_k$ is the transmitted M-PSK signal, $\rho_k$ is the fading parameter, and $n_k$ is Gaussian noise at time $k$.

The maximum *a posteriori* (MAP) algorithm can calculate the *a posteriori* probability of each bit with a perfect performance. Let $\tilde{\gamma}^{st}(s_k \rightarrow s_{k+1})$ denote the natural logarithm of the branch metric $\gamma^{st}(s_k \rightarrow s_{k+1})$, where $s_k$ is state at time $k$ and $st$ is the decoding stage, then

$$\tilde{\gamma}^{st}(s_k \rightarrow s_{k+1}) = \ln \gamma^{st}(s_k \rightarrow s_{k+1}) \tag{8.6}$$

$$\ldots = \ln P\left[d_k\right] + \ln P\left[r_k \big| x_k\right]$$

and

$$\ln P\left[d_k\right] = z_k d_k - \ln(1 + e^{z_k}) \tag{8.7}$$

$z_k$ is the *a priori* information that is obtained from the output of the other decoder. For every decoding stage of ML-TC, noisy BPSK inputs are evaluated. These noisy input sequences are obtained from the one and zero probabilities of the received signal given below.

$$P_{k,0}^{st} = \sum_{j=0}^{M/(2 \cdot st)-1} \frac{1}{(r_k - u_{0,j}^{st})^2} \tag{8.8a}$$

$$P_{k,1}^{st} = \sum_{j=0}^{M/(2 \cdot st)-1} \frac{1}{(r_k - u_{1,j}^{st})^2} \tag{8.8b}$$

$P_{k,0}^{st}$ and $P_{k,1}^{st}$ indicate zero and one probabilities of the received signal at time $k$ and stage $st$. $st$ is decoding stage and $st \in \{1, 2, \ldots, \log_2 M\}$. In an ML-TC scheme, each digit of binary correspondence of M-PSK signals matches one stage from the most significant to the least significant while stage level $st$ increases. The signal set is partitioned into the subsets because each binary digit matching stage depends on whether it is 0 or 1 as revealed in Figure 8.2. The subsets $u_{0,j}^{st}$ and $u_{1,j}^{st}$ are chosen from the signal set at stage $st$ according to the partitioning rule. With the partitioning rule, the subscripts {0,1} of $u_{0,j}^{st}$ and $u_{1,j}^{st}$ indicate the subset, which is related to the binary digit 0 or 1 at stage *st*. When the decoding stage increases, M-PSK signals are chosen among the adequate signal set in regard to the previous decoding stage as in Figure 8.2.

After computing the one and zero probabilities, as in Equations (8.8a) and (8.8b), the received signal is mapped to one-dimensional BPSK signal as below.

$$\xi_k^{st,q} = 1 - \frac{2 \cdot P_{k,0}^{st}}{P_{k,0}^{st} + P_{k,1}^{st}} \tag{8.9}$$

These probability computations and mapping are executed in every stage of the decoding process according to the signal set, as shown in Figure 8.4. Thus, Equation (8.6) becomes

$$\tilde{\gamma}^{st}(s_k \rightarrow s_{k+1}) = \ln P[d_k] - \frac{1}{2}\ln(\pi N_0 / E_s)$$

$$- \frac{E_s}{N_0} \sum_{q=0}^{n-1} \left[ \xi_k^{st,q} - (2x^q - 1) \right]^2 \tag{8.10}$$

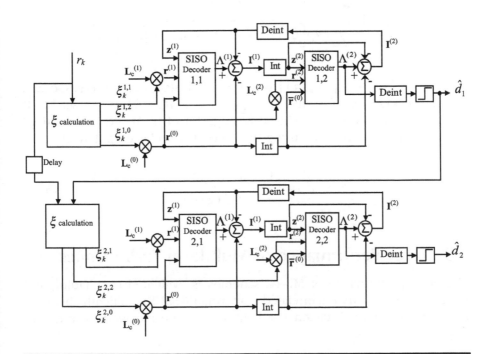

**Figure 8.4   4-PSK two-level turbo decoder structure.**

Now let $\tilde{\alpha}^{st}(s_k)$ be the natural logarithm of $\alpha^{st}(s_k)$,

$$\tilde{\alpha}^{st}(s_k) = \ln \alpha^{st}(s_k) = \ln\left\{ \sum_{s_{k-1} \in A} \exp\left[ \tilde{\alpha}^{st}(s_{k-1}) + \tilde{\gamma}^{st}(s_{k-1} \to s_k) \right] \right\} \quad (8.11)$$

where $A$ is the set of states $s_{k-1}$ connected to the state $s_k$. Now let $\tilde{\beta}^{st}(s_k)$ be the natural logarithm of $\beta^{st}(s_k)$,

$$\tilde{\beta}^{st}(s_k) = \ln \beta^{st}(s_k) \quad (8.12)$$

$$... = \ln\left\{ \sum_{s_{k+1} \in B} \exp\left[ \tilde{\beta}^{st}(s_{k+1}) + \tilde{\gamma}^{st}(s_k \to s_{k+1}) \right] \right\}$$

where $B$ is the set of states $s_{k+1}$ connected to state $s_k$, and the log likelihood ratio (LLR) can be calculated by using

$$\Lambda_k^{st} = \ln \frac{\displaystyle\sum_{S_1} \exp\left[ \tilde{\alpha}^{st}(s_k) + \tilde{\gamma}^{st}(s_k \to s_{k+1}) + \tilde{\beta}^{st}(s_{k+1}) \right]}{\displaystyle\sum_{S_0} \exp\left[ \tilde{\alpha}^{st}(s_k) + \tilde{\gamma}^{st}(s_k \to s_{k+1}) + \tilde{\beta}^{st}(s_{k+1}) \right]} \quad (8.13)$$

where $S_1 = \{ s_k \to s_{k+1} : d_k = 1 \}$ is the set of all state transitions associated with a message bit of 1, and $S_0 = \{ s_k \to s_{k+1} : d_k = 0 \}$ is the set of all state transitions associated with a message bit of 0. At the last iteration, a hard decision is done by using the second decoder output $\Lambda(2)$:

$$\hat{d}_k = \begin{cases} 1 & \text{if } \Lambda(2) \geq 0 \\ 0 & \text{if } \Lambda(2) < 0 \end{cases} \quad (8.14)$$

## 8.3 Error Performance of Multilevel Turbo Codes

The error performance of ML-TC schemes is evaluated over fading and wide-sense stationary uncorrelated scattering (WSSUS) multipath channels. In the WSSUS model, channel estimation is to be included for a severe degradation problem.

## 8.3.1 Multilevel Turbo Codes over Fading Channel

As an example, 4-PSK two-level turbo codes are simulated over AWGN, Rician, and Rayleigh channels with 100 and 256 frame sizes in Figures 8.5–8.8. ML-TC simulation results have up to 4.5 dB coding gain compared to 8-PSK turbo trellis-coded modulation (TTCM) for all cases.

Satisfactory performance is achieved in ML-TC systems for all SNR values in various fading environments. Group partitioning maximizes the Euclidean distance in these levels, and it provides an additional 0.6 dB bit error performance for all SNR. The BER and SNR curves of 4-PSK two-level turbo coding system are obtained for AWGN, Rician (for Rician channel parameter $K$ = 10 dB) and Rayleigh channels. The results are shown in Figure 8.5 and Figure 8.6 for frame sizes $N$ = 100, 256 respectively.

For comparison, error performance curves of 8-PSK turbo trellis-coded modulation (8-PSK TTCM) are obtained for the same frame sizes and channels (Figures 8.7 and 8.8).

As shown from the figures for the same channels, frame sizes, and iteration numbers, ML-TC system has up to 4.5 dB coding gain when compared to the 8-PSK TTCM system. Moreover, when 5th iteration of the 8-PSK TTCM system is compared with the first iteration of the proposed

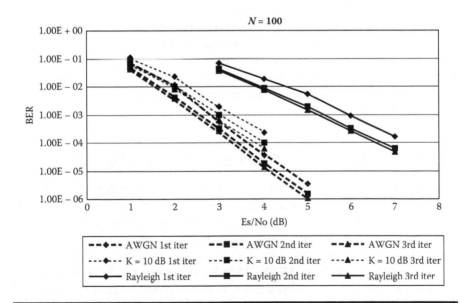

**Figure 8.5   Performance of the 4-PSK two-level turbo coding system, $N$ = 100. (Copyright permission from John Wiley & Sons [10].)**

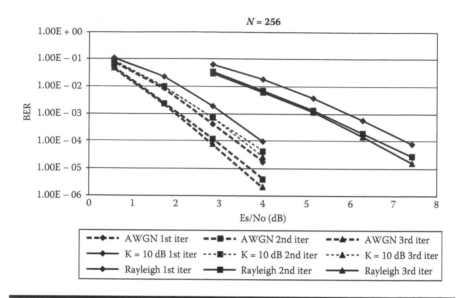

**Figure 8.6   Performance of the 4-PSK two-level turbo coding system, $N = 256$. (Copyright permission from John Wiley & Sons [10].)**

system, ML-TC has 3.3–4.1 dB coding gain. As an example, for error probability, $P_b$ of $10^{-5}$, our proposed systems have 4.5 dB coding gain with the first iteration when compared to 8-PSK TTCM with five iterations. Hence, a significant amount of coding gain is obtained without sacrificing bandwidth efficiency. The ML-TC simulation results have up to 4.5 dB coding gain compared to 8-PSK TTCM with a limited number of iterations. the desired error performance can be easily reached with a maximum of three iterations.

## 8.3.2  *Multilevel Turbo Codes over WSSUS Channels*

The performance of ML-TC schemes is evaluated over WSSUS multipath channels. In ML-TC the scheme signal set is partitioned into several levels and each level is encoded separately by the appropriate component of the turbo encoder. In the considered structure, the parallel input data sequences are encoded by our multilevel scheme and mapped to any modulation type, such as MPSK, MQAM, etc. Because WSSUS channels are very severe fading environments, it is necessary to pass the received noisy signals through nonblind or blind equalizers before the turbo decoders. In ML-TC schemes, a noisy WSSUS corrupted signal sequence is first processed in an equalizer block, then fed into the first level of the turbo decoder. The initial sequence is estimated from this first turbo decoder.

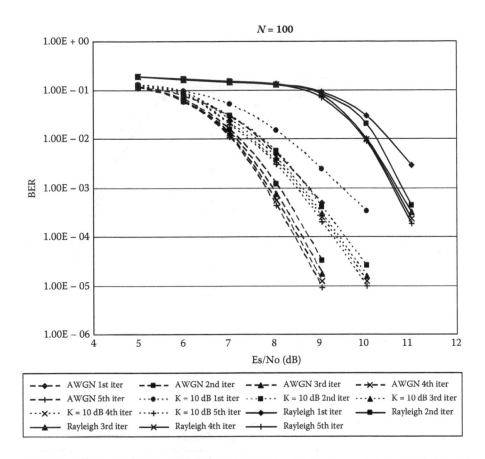

**Figure 8.7    Performance of the 8-PSK TTCM system, *N* = 100. (Copyright permission from John Wiley & Sons [10].)**

Subsequently, the following input sequences of the frame are computed by using the estimated input bit streams of previous levels.

As a ML-TC example, a 4PSK 2-level turbo codes (2L-TC) is chosen and its error performance is evaluated in the WSSUS channel modeled by COST 207 (Cooperation in the Field of Science & Technology, Project #207). It shows that 2L-TC signals with equalizer blocks exhibit considerable performance gains even at lower SNR values compared to 8PSK-TTCM [11]–[12]. The simulation results of the proposed scheme have up to 5.5 dB coding gain compared to 8PSK-TTCM for all cases. It is interesting that after a constant SNR value, 2L-TC with blind equalizer has better error performance than nonblind filtered schemes. Multilevel turbo coding maps the outputs of more than one encoder into a related symbol and each encoder is defined as a level. Because the number of encoders and decoders

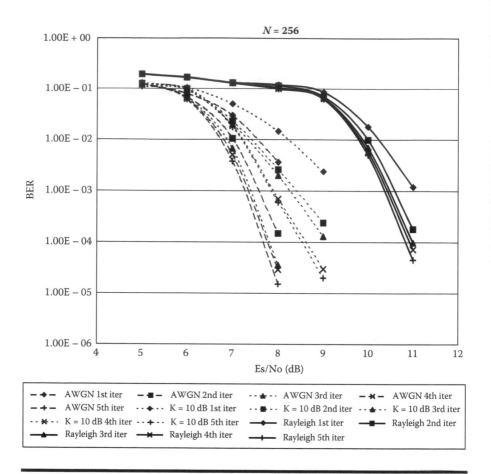

**Figure 8.8 Performance of the 8-PSK TTCM system, N = 256. (Copyright permission from John Wiley & Sons [10].)**

is the same, for each level of the multilevel encoder there is a corresponding decoder that is defined as a stage. Furthermore, except for the first decoding stage, the information bits, estimated from the previous decoding stages, are used for the next stages.

The performance of ML-TC schemes is evaluated for mobile radio communication channels, which are modeled as linear time-varying multipath channels. The simplest nondegenerate class of processes, which exhibits uncorrelated dispersiveness in propagation delay and Doppler shift, is known as the wide-sense stationary uncorrelated scattering (WSSUS) channel introduced by Bello [13]. In WSSUS, the linear superposition of uncorrelated echoes and wide-sense stationary is assumed.

Any WSSUS process is completely characterized by the two-dimensional probability density function of the propagation delays and Doppler shifts, known as scattering function. The WSSUS channel is modeled as in COST 207, and typical urban (TU), bad urban (BU), and hilly terrain (HT) profiles of standard COST 207 are chosen. Applications of the COST 207 channel include terrestrial and satellite links, such as land mobile radio, aeronautical radio, maritime radio, satellite navigation, underwater acoustical radio, and so on. Since COST 207 is a severe fading channel, receivers employ various nonblind/blind equalizers, such as least mean squares (LMS), recursive least squares (RLS), eigenvector algorithm (EVA).

## 8.3.2.1 Equalization of WSSUS Channels

In a mobile scenario, the physical multipath radio channel is time-variant with a baseband impulse response depending on the time difference $\tau$ between the observation and excitation instants as well as the (absolute) observation time $t$. Here, Equation (8.14) is rewritten to explain the adoption of the stochastic zero mean Gaussian stationary uncorrelated scattering (GSUS) model:

$$b^c(\tau, t) = \frac{1}{\sqrt{N_e}} \sum_{v=1}^{N_e} e^{j(2\pi f_{d,v} t + \Theta_v)} \cdot g_{TR}(\tau - \tau_v) \tag{8.15}$$

where $N_e$ is the number of elementary echo paths, $g_{TR}(\tau)$ denotes the combined transmit/receiver filter impulse response, and the subscript in $b^c(\cdot)$ suggests its continuous-time property. Three-dimensional sample impulse responses can easily be determined from Equation (8.15) by Doppler frequencies $f_{d,r}$ initial phases $\Theta_v$ and echo delay times $\tau_v$ from random variables with Jakes, uniform, and piecewise exponential probability density functions, respectively.

Here, the performance of multilevel turbo codes is investigated and as an example, 4PSK 2L-TC signals are transmitted over WSSUS multi path channel and corrupted by AWGN. During transmission the signals are generally corrupted due to the severe transmission conditions of WSSUS channels. These effects must be minimized to receive the signals with the fewest errors by using both equalizers and decoders. To improve performance of ML-TC signals; nonblind equalizers, such as LMS and RLS, and the blind equalizer EVA are used. Additional information about equalizers appeared in Chapter 6.

### 8.3.2.2 Simulations of ML-TC Signals over WSSUS Multipath Channel

A general block diagram of the ML-TC scheme over WSSUS channel is given below. Because WSSUS is a very severe channel model, channel equalization should be included as in Figure 8.9.

The problem decoding ML-TC codes involves the joint estimation of two Markov processes, one for each constituent code. Although in theory it is possible to model a turbo code as a single Markov process, such a representation is extremely complex and does not lend itself to computationally tractable decoding algorithms. In turbo decoding, there are two Markov processes that are defined by the same set of data. Hence, the estimation can be refined by sharing the information between the two decoders in an iterative fashion. That is, the output of one decoder can be used as *a priori* information by the other decoder. If the outputs of the individual decoders are in the form of hard-bit decisions, then there is little advantage to sharing information. However, if soft-bit decisions are produced by the individual decoders, considerable performance gains can be achieved by executing multiple iterations of decoding. The received signal can be shown as

$$r_k = \sum u_k \, b_{k-l}^c + n_k \tag{8.16}$$

where $r_k$ is noisy received signal, $u_k$ is transmitted M-PSK signal, $b_k^c$ is the slice of the time varying composite channel's impulse response and $n_k$ is Gaussian noise at time $k$.

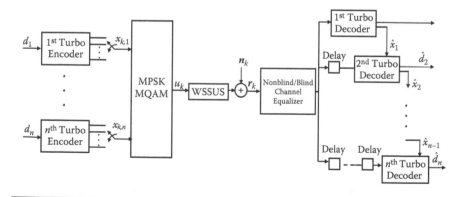

**Figure 8.9 ML-TC structure block diagram over the WSSUS channel. (Copyright permission from John Wiley & Sons [10].)**

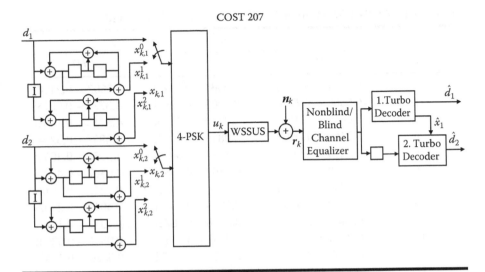

**Figure 8.10 The 2L-TC structure for WSSUS channel. (Copyright permission from John Wiley & Sons [10])**

Here, as an example of ML-TC schemes: 4PSK 2L-TC scheme with 1/3 rate RSC encoder and over all 2/3 coding rate, the generator matrix g = [111:101] and random interleavers are used as in Figure 8.10. The frame size of information symbols (bits) is chosen as $N$ = 256. Information bits are composed of data bits and training bits for nonblind (LMS and RLS) channel equalization. For blind equalization, almost all the information bits (256 bits) are used without training bits..

WSSUS channels have three main profiles. These COST 207 standards are: TU, BU, and HT. As a simulation, impulse response of $b^c(\tau = 0.25, t)$ with seventeen the channel coefficients of COST 207 (TU, BU and HT) are used as in Table 8.1 with ($\tau/T$ = 0.25) path delay

As for the echo delay time $\tau$, we use standard COST-207 TU, BU, and HT profiles. The samples of magnitude impulse responses $|b^c(\tau, t)|$, are obtained for COST 207 (TU, BU, and HT). Both time axes are normalized to the GSM symbol (bit) period $T \approx 3.7$ μs. The velocity of the mobile unit $v$ is 100 km/h. Assuming a carrier frequency of 950 MHz, this leads to a maximum Doppler shift of $f_{d,max}$ = 88 Hz. Then, Doppler period $T_{d,min} = 1/f_{d,max}$ = 3080 $T$ = 11.4 ms. Comparable statements can be made for the performance of multilevel turbo coding: For $\tau/T$ = 0.25, BER versus SNR curves of the 4PSK 2-level turbo coding system (2L-TC) are obtained for COST 207 (TU, BU and HT) by using various algorithms (RLS, LMS and EVA) as in Figures 8.11–8.13.

**Table 8.1 COST 207 Standard with 17 Channel Coefficients**

| | COST 207 | | |
|---|---|---|---|
| Path Number | TU | BU | HT |
| 1 | −0.1651 + 0.5534i | −0.0484 − 0.2943i | −1.0607 + 0.2118i |
| 2 | −0.0506 + 0.7480i | −0.0000 − 0.4282i | −1.3775 − 0.0015i |
| 3 | 0.0849 + 0.9113i | 0.0447 − 0.5339i | −1.5213 − 0.1868i |
| 4 | 0.2169 + 1.0292i | 0.0815 − 0.6063i | −1.4957 − 0.3254i |
| 5 | 0.3215 + 1.0907i | 0.1064 − 0.6414i | −1.3252 − 0.4056i |
| 6 | 0.3778 + 1.0881i | 0.1162 − 0.6368i | −1.0507 − 0.4232i |
| 7 | 0.3702 + 1.0178i | 0.1090 − 0.5919i | −0.7242 − 0.3814i |
| 8 | 0.2910 + 0.8806i | 0.0839 − 0.5077i | −0.4014 − 0.2902i |
| 9 | 0.1403 + 0.6815i | 0.0418 − 0.3872i | −0.1342 − 0.1641i |
| 10 | −0.0733 + 0.4294i | −0.0152 − 0.2351i | 0.0356 − 0.0203i |
| 11 | −0.3338 + 0.1373i | −0.0835 − 0.0577i | 0.0820 + 0.1237i |
| 12 | −0.6196 − 0.1790i | −0.1582 + 0.1375i | −0.0021 + 0.2523i |
| 13 | −0.9057 − 0.5011i | −0.2335 + 0.3418i | −0.2047 + 0.3541i |
| 14 | −1.1663 − 0.8090i | −0.3030 + 0.5460i | −0.4962 + 0.4224i |
| 15 | −1.3771 − 1.0820i | −0.3599 + 0.7407i | −0.8341 + 0.4562i |
| 16 | −1.5177 − 1.3001i | −0.3978 + 0.9165i | −1.1686 + 0.4600i |
| 17 | −1.5736 − 1.4452i | −0.4108 + 1.0649i | −1.4497 + 0.4420i |

Clearly in WSSUS channels, without equalization, it is impossible to estimate information bits. Thus an equalization block is necessary for all schemes in WSSUS channels. BER performance improvement of ML-TC schemes with equalizers increases exponentially after BER of $10^{-1}$ in all figures.

The performance of 2L-TC_RLS (2L-TC with RLS equalizer) is better than the performance of 2L-TC_LMS (2L-TC with LMS equalizer) and 2L-TC_EVA (2L-TC with EVA equalizer). In Figure 8.11, the performance of 2L-TC_LMS is better than 2L-TC_EVA except 8 dB. After 7.8 dB, the curves of 2L-TC_EVA cross over the curves of 2L-TC_LMS and approach 2L-TC_RLS.

In Figures 8.12 and 8.13, the performance of 2L-TC_EVA is better than 2L-TC_LMS at 9 dB and 12 dB, respectively. In general, equalization of 2L-TC with RLS, LMS and EVA exhibit good performance as reaching $10^{-5}$ BER values at 8 dB for COST 207 TU, $10^{-5}$ BER at 9 dB for COST 207 BU and $10^{-4}$ BER at 12 dB for COST 207 HT.

To evaluate performance of 4PSK 2L-TC, we compare our results with 8PSK-TTCM, because each of their overall coding rates is 2/3. From Figure 8.10, bit error rates of 8PSK-TTCM (no equalization) and 4PSK 2L-TC (no equalization) are quite bad at all SNR values. Thus equalization block is necessary for WSSUS environments. All types of 2L-TC have better

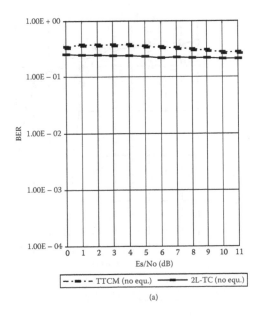

(a)

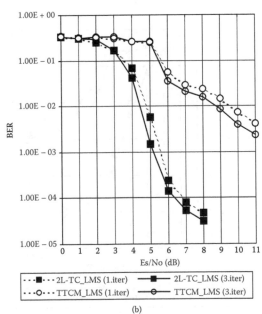

(b)

**Figure 8.11  Performance of 2L-TC and TTCM for TU type of COST 207. (a) No equalization, (b) with LMS, (c) with blind, (d) with RLS. (Copyright permission from John Wiley & Sons [11].)**

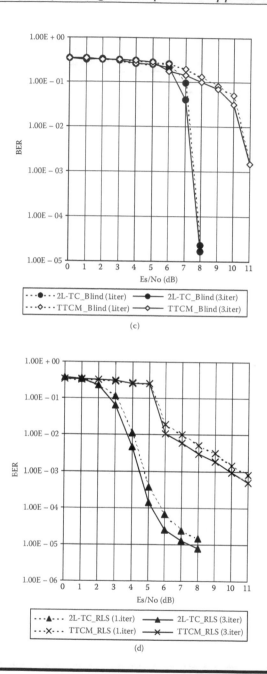

**Figure 8.11 (continued)** Performance of 2L-TC and TTCM for TU type of COST 207. **(a)** No equalization, **(b)** with LMS, **(c)** with blind, **(d)** with RLS. (Copyright permission from John Wiley & Sons [11].)

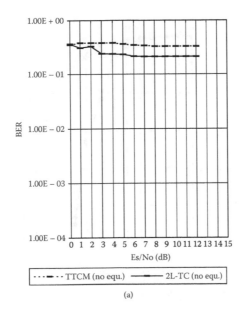

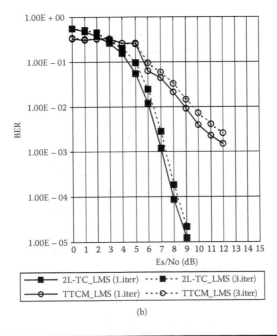

**Figure 8.12   Performance of 2L-TC and TTCM for BU type of COST 207. (a) No equalization, (b) with LMS, (c) with blind, (d) with RLS. (Copyright permission from John Wiley & Sons [11].)**

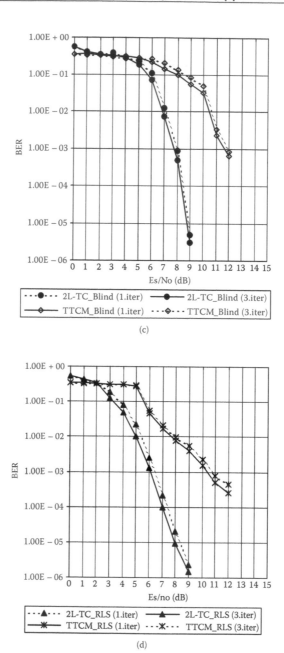

Figure 8.12 (continued)   Performance of 2L-TC and TTCM for BU type of COST 207. (a) No equalization, (b) with LMS, (c) with blind, (d) with RLS. (Copyright permission from John Wiley & Sons [11].)

error performance than TTCM after 2 dB for TU and BU and after 4 dB for HT types of COST 207. For example, in TU-type COST 207 channels, 2L-TC_RLS (3. iteration) reaches a bit error rate of $10^{-3}$ at 4.5 dB, while TTCM_RLS (3. iteration) reaches the same BER at 10 dB, thus the coding gain is about 5.5 dB (Figure 8.11). In BU type of COST 207 channels, 2L-TC_RLS (3. iteration) reaches a bit error rate of $10^{-3}$ at 6 dB, while TTCM_RLS (3. iteration) reaches the same BER at 10.5 dB, thus coding gain is about 4.5 dB (Figure 8.12). In HT-type of COST 207 channels, 2L-TC_RLS (3. iteration) reaches a bit error rate of $10^{-3}$ at 9.7 dB, while TTCM_RLS (3. iteration) reaches the same BER at 13 dB, thus coding gain is about 3.3 dB (Figure 8.13).

The performance of EVA is superior to LMS at high SNR values, because EVA has a specific feature that updates a closed-form solution iteratively similar to RLS. As mentioned in [22], after some iteration, EVA converges on the optimum linear equalizer solution and exhibits excellent convergence. For constant modulus signals, this blind approach converges as fast as the nonblind RLS algorithm. When the convergence performance of blind (EVA) is compared to nonblind (RLS); EVA uses the fourth order cumulants so it closely approaches the convergence performance of RLS for which the frame size > 140 [21]. However, it in no way approximates RLS' performance as closely as EVA does. Such equalizer structures suffer from several principal disadvantages (such as the additional time the prewhitening filter requires for convergence and the suboptimum solution in presence of additive Gaussian noise). It has been proven that, RLS will always outperform LMS [23]. As a result, the performance of EVA is close to RLS but better than LMS.

Here, multilevel turbo codes (ML-TC) are explained in detail and the performance of 2-level turbo codes (2L-TC) is studied over WSSUS channels using blind or nonblind equalization algorithms. They employ more than one turbo encoder/decoder block. Here, input bit streams are encoded and can be mapped to any modulation type. After 2L-TC modulated signals are passed through COST 207 TU, BU, and HT based on WSSUS channels. Before the decoding process, 2L-TC modulated signals are equalized blind (EVA) and nonblind algorithms (RLS and LMS). For comparison, 2L-TC signals are also sent to the receiver without equalization. At the receiver side, soft decoding is achieved in each level by using the estimated input bit streams of previous levels. Simulation results are drawn for 2L-TC and 8PSK-TTCM schemes with frame size 256. The BER of non-equalized signals (8PSK-TTCM and 2L-TC) compare poorly to equalized ones. For all three channels, the performance of 2L-TC_RLS is significantly better than 2L-TC_LMS and 2L-TC_EVA. The performance gain

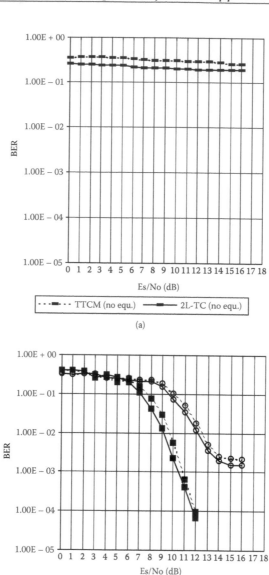

**Figure 8.13** **Performance of 2L-TC and TTCM for HT type of COST 207. (a) No equalization, (b) with LMS, (c) with blind, (d) with RLS. (Copyright permission from John Wiley & Sons [11].)**

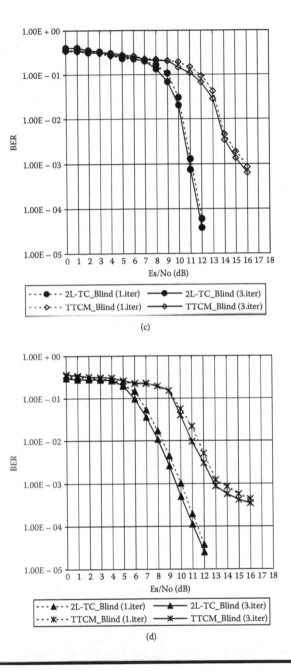

Figure 8.13 (continued)  Performance of 2L-TC and TTCM for HT type of COST 207. (a) No equalization, (b) with LMS, (c) with blind, (d) with RLS. (Copyright permission from John Wiley & Sons [11].)

of 2L-TC_LMS over 2L-TC_EVA is better for lower values of SNR, but after certain values of SNR (8 dB for TU, 9 dB for BU, and 12 dB for HT), 2L-TC_EVA outperforms 2L-TC_LMS. It is obvious that the application of 2L-TC with equalization over WSSUS multipath exhibits a good performance at lower SNR values. Thus ML-TC schemes can be considered as a compromising approach for WSSUS channels if they are accompanied by equalization blocks.

# References

[1] Berrou, C., A. Glavieux, and P. Thitimasjshima. Near Shannon-Limit Error Correcting Coding and Decoding: Turbo Codes (1). *Proc., IEEE Int. Conf. on Commun.* (Geneva, Switzerland):1064–1070, May 1993.

[2] Robertson, P., and T. Worzt. Bandwidth-Efficient Turbo Trellis-Coded Modulation Using Punctured Component Codes. *IEEE J. SAC* 16, Feb. 1998.

[3] Blackert, W., and S. Wilson S. Turbo Trellis Coded Modulation. *Conference Information Signals and System, CISS'96.*

[4] Osman, O., O. N. U•an O.N. Blind Equalization of Turbo Trellis-Coded Partial-Response Continuous-Phase Modulation Signaling over Narrow-Band Rician Fading Channels. *IEEE Transactions on Wireless Communications* 4(2):397–403, Mar. 2005.

[5] U•an, O. N. Jitter and Error Performance of Trellis Coded Systems over Rician Fading Channels with Imperfect Phase Reference. *Electronic Letter* 32:1164–1166, 1996.

[6] Imai, H., and S. Hirakawa. A New Multilevel Coding Method Using Error-Correcting Codes. *IEEE Trans. on Inform. Theory* IT-23:371–377, May 1977.

[7] Yamaguchi, K., and H. Imai. Highly Reliable Multilevel Channel Coding System Using Binary Convolutional Codes. *Electron. Lett.*, 23(18):939–941, Aug.1987.

[8] Pottie, G. J., and D. P.Taylor. Multilevel Codes based on Partitioning. *IEEE Trans. Inform. Theory* 35:87–98, Jan. 1989.

[9] Calderbank, A. R. Multilevel Codes and Multistage Decoding. *IEEE Trans. Commun.* 37:222–229, Mar. 1989.

[10] Osman, O., O. N. U•an, and N. Odabasioglu. Performance of Multi Level-Turbo Codes with Group Partitioning over Satellite Channels. *IEEE Proc. Communications* 152(6):1055–1059, 2005.

[11] U•an, O. N., K. Buyukatak, E. Gose, O. Osman, and N. Odabasioglu. Performance of Multilevel-Turbo Codes with Blind/Non-Blind Equalization over WSSUS Multipath Channels. *Int. Journal of Commun. Syst.* 19:281–297, 2006.

[12] Osman, O. Blind Equalization of Multilevel Turbo Coded-Continuous Phase Frequency Shift Keying over MIMO Channels. *Int. Journal of Commun. Syst.* 20(1):103–119, 2007.

[13] Bello, P. A. Characterization of Randomly Time-Variant Linear Channels. *IEEE Transactions on Communication Systems* CS-11:360–393, Dec. 1963.

[14] Boss, D., B. Jelonnek, and K. Kammeyer. Eigenvector Algorithm for Blind MA System Identification. *Elsevier Signal Processing* 66(1):1–26, April 1998.

[15] Hoeher, P. A Statistical Discrete-Time Model for the WSSUS Multipath Channel. *IEEE Trans. on Vehicular Technology* VT-41(4):461–468, April 1992.

[16] Boss, D., B. Jelonnek, and K. Kammeyer. Eigenvector Algorithm for Blind MA System Identification. *Elsevier Signal Processing* 66(1):1–26, April 1998.

[17] Shalvi, O. and E. Weinstein. New Criteria for Blind Deconvolution of Nonminimum Phase Systems (Channels). *IEEE Trans. on Information Theory* IT-36(2):312–321, Mar. 1990.

[18] Shalvi, O., and E. Weinstein. Super-Exponential Methods for Blind Deconvolution. *IEEE Trans. on Information Theory* 39(2):504–519, Mar. 1993.

[19] Jelonnek, B. and K. D. Kammeyer. Closed-Form Solution to Blind Equalization. *Elsevier Signal Processing, Special Issue on Higher Order Statistics* 36(3):251–259, April 1994.

[20] Stüber, G. L. *Principles of Mobile Communication.* Norwell, MA: Kluwer Academic Publishers, 2001.

[21] Jelonnek, B., D. Boss, and K. D. Kammeyer. Generalized Eigenvector Algorithm for Blind Equalization. *Elsevier Signal Processing* 61(3):237–264, Sept. 1997.

[22] Jelonnek, B. and K. D. Kammeyer. A Blind Adaptive Equalizer based on Lattice/All Pass Configuration. *In Proc. EUSIPCO-92* II:1109–1112, Brussels, Belgium, Aug. 1992.

[23] Wang, R., N. Jindal, T. Bruns, A. R. S. Bahai, and D. C. Cox. Comparing RLS and LMS Adaptive Equalizers for Nonstationary Wireless Channels in Mobile ad hoc Networks. *Proceedings of IEEE International Symposium on Personal, Indoor and Mobile Radio Communications* 3:1131–1135, Lisbon, Portugal, 2002.

# Chapter 9

# Turbo Trellis-Coded Modulation

Turbo codes are a new class of error correction codes that were introduced along with a practical decoding algorithm in [1]. Berrou et al. introduced turbo codes in 1993, which achieved high performance close to the Shannon limit. Then, turbo codes were suggested for low-power applications such as deep-space and satellite communications, as well as for interference-limited applications such as third generation cellular and personal communication services.

Blackert and Wilson concatenated turbo coding and trellis-coded modulation for multilevel signals in 1996. As a result, turbo trellis-coded (TTC) modulation was developed as in [6]. The main principle of turbo codes was applied to TCM by retaining the important properties and advantages of both of their structures as in [2,6]. TCM has a systematic-feedback convolutional encoder followed by a signal mapper for one or more dimensional codes. In turbo trellis-coded modulation (TTCM), two parallel binary recursive convolutional encoders are used and interleaving, deinterleaving, and puncturing are adapted as placed in [2]. TTCM is a bandwidth-efficient channel coding scheme for multidimensional input symbols that has a structure similar to binary turbo codes, but employs trellis-coded modulation.

## 9.1 Turbo Trellis Encoder

The most important characteristic of turbo codes is their simple use of recursive-systematic component codes in a parallel concatenation scheme.

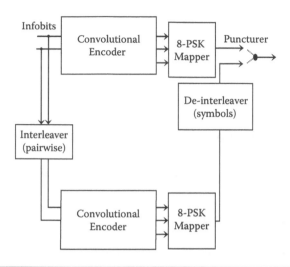

**Figure 9.1   TTCM encoder structure.**

Pseudorandom bit-wise interleaving between encoders ensures a small bit-error probability [2,6]. In [2], Ungerboeck codes (and multidimensional TCM codes) have been employed as building blocks in a turbo coding scheme in a similar way as binary codes were used [1].

The schema of an encoder in Figure 9.1 was presented in [2,4,5]. As an example of simulation, $\tilde{m} = m = 2$ ($\tilde{m}$ = coded input bits, $m$ = total input bits), $N = 8$ ($N$ = frame size) are chosen for 8-PSK signaling. The mapper structure is shown in Figure 9.2. The two-bit input stream includes an eight-symbol sequence $(d_1, d_2, \ldots, d_8) = (00, 11, 10, 01, 11, 10, 01, 00)$. These information bit pairs are encoded in an Ungerboeck-style encoder to yield the 8-PSK sequence (0, 6, 5, 3, 6, 4, 2, 0). The information bits are interleaved on a pairwise basis according to (3 6 4 7 2 1 8 5) and (10, 11, 00, 10, 00, 11, 01, 01) sequence is obtained and encoded again into the sequence (4, 6, 0, 5, 0, 7, 2, 2). The output symbols of the second encoder are deinterleaved to guarantee that the ordering of the two information bits partly defining each symbol corresponds to that of the first encoder. The output sequence of the deinterleaver is (0, 7, 4, 2, 6, 5, 2, 0). Finally, puncturer allows the first symbol to pass from the first encoder, the second symbol from the second encoder, and so forth. The output sequence of the TTCM is (0, 7, 5, 2, 6, 5, 2, 0).

By using interleaver and deinterleaver for the second encoder, the symbol index $k$ before the interleaver is associated with the same indexed symbol at the deinterleaver output [2]. The interleaver must map even positions to even positions and odd positions to odd ones [2]. For any modulation type and any $m$ number, each information bit influences either the upper or lower encoder's state but never both.

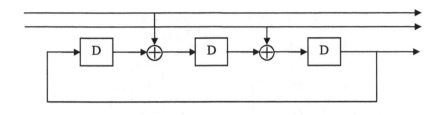

**Figure 9.2    2/3 Convolutional encoder with three memories.**

A convolutional encoder with two inputs, three outputs, and three memories is shown in Figure 9.2.

## 9.2 Turbo Trellis Decoder

The turbo trellis decoder and binary turbo decoder are very similar. One important difference, which affects the decoding algorithm, is that in binary turbo coding, systematic and parity data are sent separately; but in TTCM, systematic and parity bits are mapped into the same symbol. Therefore, in binary turbo coding, the decoder's output can be split into three parts. These parts are a systematic component, an *a priori* component, and an extrinsic component. But only the extrinsic component may be given to the next decoder; otherwise, information will be used more than once in the next decoder [1].

Here, for the turbo trellis decoder the systematic information is transmitted with parity information in the same symbol. However, the decoder output can be split into two different components—the first one is *a priori* and the second is extrinsic and systematic together [2].

In Figure 9.3, an asterisk (*) shows the switch position when the even-indexed signals pass. Here, metric *s* calculation is only used in the first decoding step. Because of the structure and modulation of the TTCM encoder, systematic bits and parity bits are mapped to one signal, while at the receiver, parity data cannot be separated from the received signal. Hence, in the first half of the first decoding for the symbol-by-symbol (S-b-S) MAP decoder-1, *a priori* information cannot be generated. Metrics calculates the *a priori* information from the even index in the received sequence.

### 9.2.1  Metric Calculation

Metric *s* calculation is used in the very first decoding stage. If the first decoder sees a group of *n* punctured symbols in the first decoding stage, there is no *a priori* input. Therefore, before the first decoding can pass

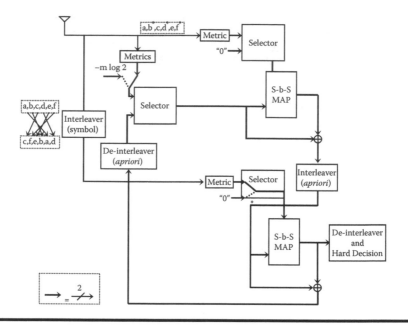

**Figure 9.3  TTCM decoder.**

off the upper decoder, the *a priori* information to contain the systematic information for the * transitions is set, where the symbol was transmitted partly by the information group $d_k$, but also the unknown parity bit $b^{0,*}$ $\in \{0,1\}$ produced by the other encoder [2].The *a priori* information is set by applying the mixed Bayes' rule as in [2]:

$$\Pr\{d_k = i \mid y_k\} = const \cdot p(\mathbf{y_k} \mid d_k = i)$$

$$= const \cdot \sum_{j \in \{0,1\}} p(\mathbf{y_k}, b_k^{0,*} = j \mid d_k = i) \qquad (9.1)$$

$$= \frac{const}{2} \cdot \sum_{j \in \{0,1\}} p(\mathbf{y_k} \mid d_k = i, b_k^{0,*} = j)$$

where it is assumed that $\Pr\{b_k^{0,*} = j \mid d_k\} = \Pr\{b_k^{0,*} = j\} = 1/2$ and $\mathbf{y_k} = (y_k^0, \ldots, y_k^{(n-1)})$ if the receiver observes $N$ set of $n$ noisy symbols, where $n$ such symbols are associated with each step in the trellis. In Equation (9.1), the value of the constant is not calculated because the value of $\Pr\{d_k = i \mid \mathbf{y_k}\}$ is found by normalization. This normalization can be achieved by dividing the summation

$$\sum_{j \in \{0,1\}}$$

by its sum over all $i$ [2]. If the upper decoder is not at a * transition, then $\Pr\{d_k = i\}$ is set to $1/2^m$ [2].

## 9.2.2 MAP Decoder

To begin, the state transitions must be calculated by the given formulation below [2].

$$\gamma_i(y_k, M', M) = p(y_k \mid d_k = i, S_k = M, S_{k-1} = M')$$

$$\cdot q(d_k = i \mid S_k = M, S_{k-1} = M') \qquad (9.2)$$

$$\cdot \Pr\{S_k = M \mid S_{k-1} = M'\}$$

$q(d_k = i \mid S_k = M, S_{k-1} = M')$ is either zero or one, depending on whether encoder input $i \in \{0,1, \dots, 2^m-1\}$ is associated with the transition from state $S_{k-1} = M'$ to $S_k = M$ or not. In the last component of Equation 9.2, *a priori* information is as [2]

$$\Pr\{S_k = M \mid S_{k-1} = M'\}$$

$$= \begin{cases} \Pr\{d_k = 0\}, & if\ q(d_k = 0 \mid S_k = M, S_{k-1} = M') = 1 \\ \Pr\{d_k = 1\}, & if\ q(d_k = 1 \mid S_k = M, S_{k-1} = M') = 1 \\ \quad\vdots \\ \Pr\{d_k = 2^m - 1\}, if\ q(d_k = 2^m - 1 \mid S_k = M, S_{k-1} = M') = 1 \end{cases} \qquad (9.3)$$

$$= \Pr\{d_k = j\}$$

where $j: q(d_k = j \mid S_k = M, S_{k-1} = M') = 1$. If there is no such $j$, $\Pr\{S_k = M, S_{k-1} = M'\}$ is set to zero.

The calculations are less complex in the log domain, thus, the Jacobian logarithm is considered.

$$\ln(e^x + e^y) = \max(x, y) + \ln(1 + e^{-|y-x|})$$

$$= \max(x, y) + f_c(|x - y|) \qquad (9.4)$$

Now let $\tilde{\gamma}_i(s_k \to s_{k+1})$ be the natural logarithm of $\gamma_i(s_k \to s_{k+1})$

$$\tilde{\gamma}_i(s_k \to s_{k+1}) = \ln \gamma_i(s_k \to s_{k+1}) \qquad (9.5)$$

Now let $\tilde{\alpha}(s_k)$ be the natural logarithm of $\alpha(s_k)$,

$$= \ln \left\{ \sum_{s_{k-1} \in A} \exp\left[ \tilde{\alpha}(s_{k-1}) + \tilde{\gamma}_i(s_{k-1} \rightarrow s_k) \right] \right\} \tag{9.6}$$

where $A$ is the set of states $s_{k-1}$ that are connected to the state $s_k$. Now, let $\tilde{\beta}(s_k)$ be the natural logarithm of $\beta(s_k)$:

$$\tilde{\beta}(s_k) = \ln \beta(s_k)$$

$$= \ln \left\{ \sum_{s_{k+1} \in B} \exp\left[ \tilde{\beta}(s_{k+1}) + \tilde{\gamma}_i(s_k \rightarrow s_{k+1}) \right] \right\} \tag{9.7}$$

where $B$ is the set of states $s_{k+1}$ that are connected to state $s_k$. Therefore, the desired output of the MAP decoder is

$$P_r\{d_k = i \mid \underline{\mathbf{y}}\} = const \cdot \sum_M \sum_{M'} \left[ \tilde{\gamma}_i(y_k, M', M) + \tilde{\alpha}_{k-1}(M') + \tilde{\beta}_k(M) \right] \tag{9.8}$$

$\forall \; i \in \{0, \dots, 2^{m-1}\}$. The constant can be eliminated by normalizing the sum of above formulation over all $i$.

## 9.3 Error Performance of Turbo Trellis-Coded Modulation

As an example, the error performance curves of 8-PSK TTCM are obtained for frame sizes $N = 100$ and $256$.

In Figure 9.4, as iteration number increases, error performance improves. In Rician channels, the fading parameter is chosen as 10 dB.

In both Figures 9.4 and 9.5, BER improves as SNR increases. Frame size $N$ is one of the main parameters in performance analysis.

## 9.4 Time Diversity Turbo Trellis-Coded Modulation (TD-TTCM)

As previously stated, turbo codes are the most efficient codes for low-power applications, such as deep-space and satellite communications, as well as for interference-limited applications, such as third generation

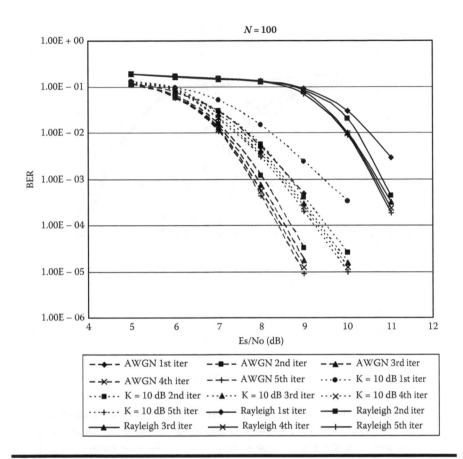

**Figure 9.4    Error performance of 8-PSK TTCM for various channels for $N$ = 100.**

cellular and personal communication services. Because turbo codes use convolutional codes as their constituent codes, a natural extension of the turbo concept that improve bandwidth efficiency, they can be systems using TCM. As in [1], the main principle of turbo codes is applied to TCM by retaining the important properties and advantages of both structures. TCM uses two parallel recursive encoders as in [2,6] like turbo codes, but the input-bit streams are more than one.

To improve error performance and time diversity turbo trellis coded modulation schemes are combined and denoted as (TD-TTCM) [7]. The general block diagram of the considered scheme is given in Figure 9.6.

Here, performance of TD-TTCM is evaluated for mobile radio communication channels that are modeled as linear time-varying multipath MIMO channels. In Multi Input Multi Output (MIMO) transmission, high

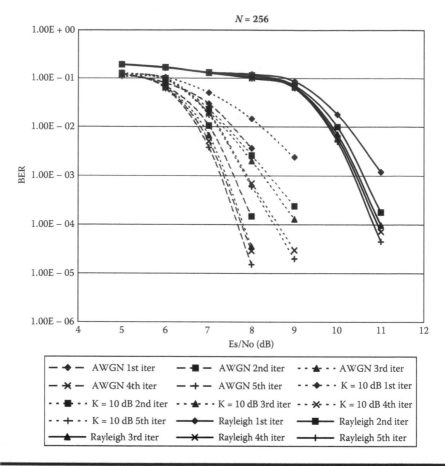

**Figure 9.5    Error performance of 8-PSK TTCM for various channels for *N* = 256.**

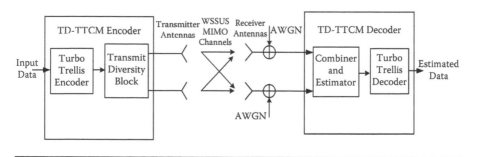

**Figure 9.6    General block diagram of TD-TTCM [7].**

data rates are achieved by multiple antennas at the transmitter and receiver, opening equivalent parallel transmission channels [7–9,10]. The improved reliability of communication results from increased transmitter or receiver diversity.

In the TD-TTCM scheme, input binary data is passed through a turbo trellis encoder followed by a transmit diversity block shown in Figure 9.6. At the transmitter side, a number of antennas are placed separately to achieve transmit diversity. Then the TD-TTCM encoded data stream is disturbed by a severe noisy COST 207–type communication environment that is modeled by WSSUS with the MIMO channel. The simplest nondegenerate class of processes that exhibits uncorrelated dispersiveness in propagation delay and Doppler shift is known as the WSSUS channel, introduced by Bello [12]. The WSSUS channel is modeled as COST 207, TU, BU, and HT. Because of the severity of the fading channel, it is necessary to employ a powerful channel estimator, such as a genetic algorithm (GA) at the receiver side. After transmission, the distorted multipath signals arrive at a number of receiver antennas. At the decoder side, the turbo trellis decoder follows the genetic-based estimator and the combiner, which carry iterative solutions. The main blocks of the TD-TTCM structure are explained in detail as follows.

## 9.4.1  TD-TTCM Encoder

In the TD-TTCM encoder scheme, the time diversity block is placed after the classical TTCM encoder structure. In Figure 9.7, TD-TTCM is drawn for 8-PSK, where input binary data is passed through a turbo trellis encoder followed by a transmit diversity block.

## 9.4.2  TD-TTCM Decoder

The turbo trellis decoder is similar to that used to decode binary turbo codes, except for a difference passed from one decoder to the other, and the treatment of the very first decoding step. In binary turbo coding, the decoder's output can be split into three parts. These parts are a systematic component, an *a priori* component, and an extrinsic component [6]. But only the extrinsic component may be given to the next decoder; otherwise, information will be used more than once in the next decoder [1]. Here, for the turbo trellis decoder the systematic information is transmitted with parity information in the same symbol. However, decoder output can be

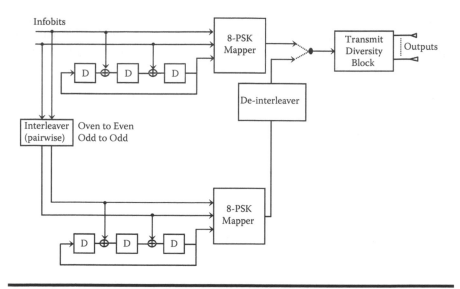

**Figure 9.7    Encoder of TD-TTCM for 8-PSK [7].**

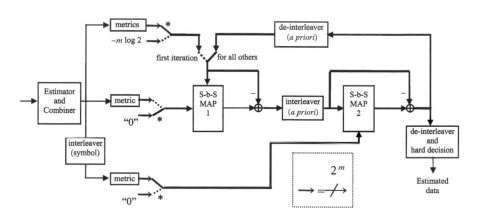

**Figure 9.8    TD-TTCM decoder [7].**

split into two different components, first one is *a priori* and the second is extrinsic and systematic together (Figure 9.8).

In Figure 9.8, the structure and the decoding algorithm of TD-TTCM and TTCM is the same except for the estimator and combiner block.

## 9.4.3 WSSUS MIMO Channel Model of the TD-TTCM Scheme

The discrete-time system model and the complex baseband signaling using transmit antennas and receive antennas of a WSSUS MIMIO channel is shown in Figure 9.9.

At time $k$, the vector symbol $S[k] \equiv [s_1[k] \ldots s_{N_T}[k]]^T$ is composed of the entries of the vector symbols. $\mathbf{s}[k]$ is transmitted over the WSSUS MIMO channel denoted as $\mathbf{H}[k]$, where $\mathbf{H}[k]$ is an $n_T \times n_R$ matrix with $\mathbf{h}^{\mu\nu}[k]$ entry, $1 \leq \mu \leq n_R$, $1 \leq \nu \leq n_T$, in the $\mu$th row and $\nu$-th column accounting for WSSUS channel coefficients between transmit antenna $\nu$ and receive antenna $\mu$. The symbols $\mathbf{s}_\nu[k]$, $1 \leq \nu \leq n_T$ are drawn from an M-ary alphabet, e.g., a PSK or QAM constellation, with equal transmit power for each antenna [3,17]. Collecting the received samples in a vector $y[k] \equiv [y_1[k] \ldots y_{N_T}[k]]^T$, the input–output relation becomes

$$\mathbf{y}[k] = \mathbf{H}[k]\mathbf{s}[k] + \mathbf{n}[k] \qquad (9.9)$$

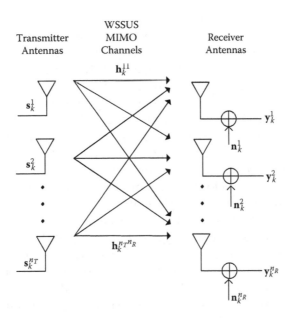

**Figure 9.9   WSSUS MIMO channels.**

where $n[k] \equiv [n_1[k] \ldots n_{N_T}[k]]^T$ denotes the complex AWGN assumed to be spatially and temporally uncorrelated and with equal variance,

$$\sigma_n^2 = \left\{ \left| \mathbf{n}_\mu[k] \right|^2 \right\},$$

at each receiver antenna.

The model that is used here is a physical multipath radio channel with time-variant employing a base band impulse response depending on the time difference $\tau$ between the observation and excitation instants as well as the (absolute) observation time $k$. The stochastic zero mean Gaussian stationary uncorrelated scattering (GSUS) model should be adopted leading to the following impulse response of the composite channel [13,14]

$$h(\tau, k) = \frac{1}{\sqrt{N_e}} \sum_{v=1}^{N_e} e^{j(2\pi f_{d,v} k + \Theta_v)} \cdot g_{TR}(\tau - \tau_v) \qquad (9.10)$$

where $N_e$ is the number of elementary echo paths, $g_{TR}(\tau)$ denotes the combined transmit/receive filter impulse response, and the subscript in $h(\cdot)$ suggests its continuous-time property. 3D sample impulse responses can easily be determined from Equation (9.10) by Doppler frequencies $f_{d,v}$ initial phases $\Theta_v$ and echo delay times $\tau_v$ from random variables with Jakes, uniform, and piecewise exponential probability density functions, respectively. As for the echo delay time $\tau$, use the standard COST-207 TU, BU, and HT profiles. The samples of magnitude impulse responses $|h(\tau, k)|$ of COST 207 (TU, BU, and HT) obtained from Equation (9.10). In the COST 207 model, both time axes are normalized to the GSM symbol period $T \approx 3.7$ μs. The velocity of the mobile unit is $v = 100$ km/h. Assuming a carrier frequency of 950 MHz, this leads to a maximum Doppler shift of $f_{d,max} = 88$ Hz. Equation (9.10) is evaluated over a $k$ range covering one minimum Doppler period $T_{d,min} = 1/f_{d,max} = 3080T = 11.4$ ms.

The genetic algorithm (GA) is a good choice for observing optimal solution in TD scheme. In nonlinear channel models, GA is one of the most powerful tools for searching global solutions. The other methods that search local solutions may not always find the possible optimal channel parameters. But to obtain an optimal solution, a transmit diversity scheme (repetition code) is necessary to equalize the number of unknown parameters and equations at the combiner on the receiver side. The transmitted data must be sent in repetition form, to estimate the symbol vectors that are sent from the transmitter antennas. TD also has a great advantage compared to classical space time (ST) coding, because transmission in ST is a symbol-by-symbol form, but in TD the transmission and solution are in vector form. In classic ST schemes, fading channel parameters keep constant for consecutive symbols, while in our TD model, these parameters may be different for each symbol, thus it simulates the channel as modeled in real life by COST 207 standards [14].

### 9.4.3.1 Transmit Diversity Scheme

The TD-TTCM coded signals are transmitted by using a transmit diversity block as is shown in Figure 9.10. At the $k$th coding step input of a transmit diversity block is a block symbol vectors, each vector $S[k] \equiv [s_1[k] \ldots s_{N_T}[k]]^T$ involves $M$ complex symbols $\{x_k + jy_k \quad k = 1, \cdots, M\}$, where $x_k + jy_k$ are elements of a higher order modulation constellation, e.g., *M-PSK*.

The TD scheme improves the signal by simple processing across two transmitter antennas and one receiver antenna as shown in Figure 9.10. TTCM modulated 16 signals are simultaneously transmitted from the two transmitter antennas, $Tx_1$ and $Tx_2$. In the first coding step, the signals transmitted from $Tx_1$ are denoted as $\mathbf{s}_1 = [x_1 + jy_1 \quad x_2 + jy_2 \ldots x_8 + jy_8]$, and those from $Tx_2$ are denoted as $\mathbf{s}_2 = [x_9 + jy_9 \quad x_{10} + jy_{10} \ldots x_{16} + jy_{16}]$. As

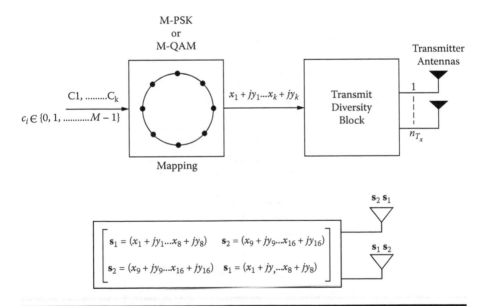

**Figure 9.10  (a) Mapping and transmit diversity (TD) block, (b) transmission of symbol vectors by TD [7].**

shown in Figure 9.10, in the second coding step; antenna $Tx_1$ transmits $\mathbf{s}_2$, while antenna $Tx_2$ transmits $\mathbf{s}_1$. Assume that WSSUS channel parameters are constant during these two consecutive symbol vectors. The channel parameters are between $Tx_1$ and receiver antenna $(Rx)$ are defined as $\mathbf{h}^{11} = [h_1 + jz_1 \quad h_2 + jz_2 \ldots h_8 + jz_8]$. The channel between the transmitter antenna, $Tx_2$ and $Rx$ are defined as $\mathbf{h}^{21} = [h_9 + jz_9 \ h_{10} + jz_{10} \ldots h_{16} + jz_{16}]$. 2Tx-1Rx (2 transmitters, 1 receiver) configured TD-TTCM decoder is shown in Figure 9.11.

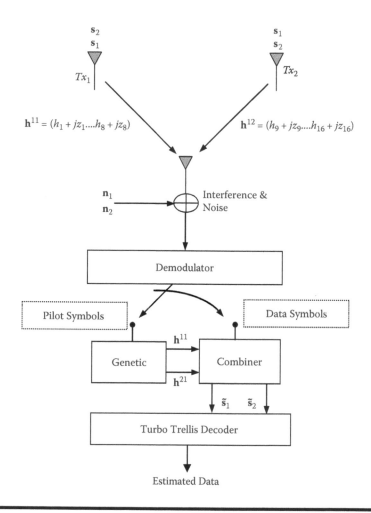

**Figure 9.11   Block diagram of decoder of TD-TTCM for 2Tx-1Rx antenna configuration [7].**

The multipath channels are all complex and modeled as specific channels (TU, BU, and HT) of COST 207 [13,14], as below

$$\mathbf{h}^{11}(k) = \mathbf{h}^{11}(k + T) = [h_1 + jz_1 \square h_8 + jz_8]$$

$$\mathbf{h}^{21}(k) = \mathbf{h}^{21}(k + T) = [h_9 + jz_9 \square h_{16} + jz_{16}] \qquad (9.11)$$

where $T$ is the symbol duration. Then the received signals can be expressed as,

$$\mathbf{y}_1 = \mathbf{y}(k) = \mathbf{h}^{11} * \mathbf{s}_1 + \mathbf{h}^{21} * \mathbf{s}_2 + \mathbf{n}_1$$

$$\mathbf{y}_2 = \mathbf{y}(k + T) = \mathbf{h}^{11} * \mathbf{s}_2 + \mathbf{h}^{21} * \mathbf{s}_1 + \mathbf{n}_2 \qquad (9.12)$$

where $\mathbf{y}_1$ and $\mathbf{y}_2$ are received signals at time $k$ and $k + T$, and $\mathbf{n}_1$ and $\mathbf{n}_2$ are Gaussian noise with the noise variance is

$$\sigma^2 = \frac{N_0}{2E_s}$$

and $*$ is convolution operator.

## 9.4.3.2 Genetic Algorithm (GA) and Genetic-Based Channel Estimator

GAs are search and optimization algorithms based on the principle of natural evolution and population genetics. GAs were initially proposed by Holland [15]. They have been applied successfully to many engineering and optimization problems. GAs use different genetic operators to manipulate individuals in a population over several generations to improve their fitness gradually. GAs maintain a set of candidate solutions called a population. Candidate solutions are usually represented as strings called chromosomes, coded with a binary character set {0,1}. The GA has a population of individuals. Each individual represents a potential solution to the problem in question. Individuals are evaluated to give a measure of their fitness.

To involve the best solution candidate (or chromosome), GA employs the genetic operators of selection, crossover, and mutation for manipulating the chromosomes in a population. The flow chart of a simple GA is shown in Figure 9.12.

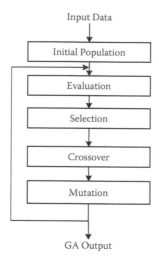

**Figure 9.12  Flowchart for a simple genetic algorithm.**

In the selection step, individuals are chosen to participate in the reproduction of new individuals. Different selection methods, such as stochastic selection or ranking-based selection, can be used. The crossover operator combines characteristics of two parent individuals to form two offspring. Mutation is a random alteration with a small probability of the binary value of a string position. This operation will prevent the GA from being trapped in a local minimum. The fitness-evaluation unit in the flow chart acts as an interface between the GA and the optimization problem. Information generated by this unit about the quality of different solutions is used by the selection operation in the GA. More details about GAs can be found in [16].

At the genetic-based channel estimator, $\mathbf{y}(k)$ is the output vector at time $k$. The GA will be used to search for channel coefficients $\mathbf{h}(k)$ that will minimize the square root of error between the output vector $\mathbf{y}(k)$ and $\mathbf{s}(k) * \mathbf{h}(k)$ at time $k$. The following performance index can be defined

$$\mathbf{y}(k) = \mathbf{s}(k) * \mathbf{h}(k) + \mathbf{n}(k) \tag{9.13}$$

$$\mathbf{J}(k) = \sqrt{\mathbf{y}(k) - \mathbf{s}(k) * \mathbf{h}(k)} \tag{9.14}$$

A genetic-based channel estimator searches for the optimal value of channel coefficients $\mathbf{h}(k)$ that will minimize $\mathbf{J}$. The following steps describe the operation of the proposed GA-based estimator. At time step $k$:

- Evaluate $\mathbf{y}(k)$ using WSSUS multi path channel model.
- Use GA search to find $\mathbf{h}(k)$ which will minimize the performance index $\mathbf{J}$. WSSUS channel model should be used to find $\mathbf{y}(k)$ for different values of $\mathbf{h}(k)$.
- Apply the optimal value of $\mathbf{h}(k)$ generated in the previous step to $\mathbf{y}(k)$.
- Repeat for time $k + 1$.

### 9.4.3.3 The Structure of the Combiner

As an example, we consider the combiner in Figure 9.11, and the output symbols of the combiner are fed to the TD-TTCM decoder. In the combiner, symbols are extracted from $\mathbf{y}_1$ and $\mathbf{y}_2$ by using the iterative solution. Sixteen symbols ($\mathbf{s}_1$ and $\mathbf{s}_2$) that are transmitted by both antennas passed through 16 WSSUS channels ($\mathbf{h}^{11}$ and $\mathbf{h}^{21}$) and AWGN. After one symbol vector duration, 16 symbols are also transmitted. After the convolution of the signal vector with the channel vector for each antenna, finally we conclude 32 equations and 32 unknown parameters $(\tilde{x}_1, \tilde{y}_1, \tilde{x}_2, \tilde{y}_2, \square, \tilde{x}_{16}, \tilde{y}_{16})$. The matrix form of the 32 equations is as [7,8]

$$\begin{bmatrix} \tilde{x}_1 \\ \tilde{y}_1 \\ \tilde{x}_9 \\ \tilde{y}_9 \end{bmatrix} = [\tilde{\mathbf{a}}]^{-1} * \begin{bmatrix} A \\ B \\ C \\ D \end{bmatrix} \quad \text{where } \tilde{\mathbf{a}} = \begin{bmatrix} b_1 & -z_1 & b_9 & -z_9 \\ z_1 & b_1 & z_9 & b_9 \\ b_9 & -z_9 & b_1 & -z_1 \\ z_9 & b_9 & z_1 & b_1 \end{bmatrix} \quad (9.15)$$

From Equation (9.15), the first symbol of each transmitter antenna is extracted. Then we put $\tilde{x}_1, \tilde{y}_1, \tilde{x}_9, \tilde{y}_9$ into the following equation:

$$\begin{bmatrix} \tilde{x}_2 \\ \tilde{y}_2 \\ \tilde{x}_{10} \\ \tilde{y}_{10} \end{bmatrix} = [\tilde{\mathbf{a}}]^{-1} * \left\{ \begin{bmatrix} E \\ F \\ G \\ H \end{bmatrix} - [\tilde{\mathbf{b}}] * \begin{bmatrix} \tilde{x}_1 \\ \tilde{y}_1 \\ \tilde{x}_9 \\ \tilde{y}_9 \end{bmatrix} \right\}$$

where

$$\tilde{\mathbf{b}} = \begin{bmatrix} b_2 & -z_2 & b_{10} & -z_{10} \\ z_2 & b_2 & z_{10} & b_{10} \\ b_{10} & -z_{10} & b_2 & -z_2 \\ z_{10} & b_{10} & z_2 & b_2 \end{bmatrix} \quad (9.16)$$

$\{\tilde{x}_2, \tilde{y}_2, \tilde{x}_{10}, \tilde{y}_{10}\}$ are extracted from Equation (9.16). In the same manner,

$$
\begin{bmatrix} \tilde{x}_3 \\ \tilde{y}_3 \\ \tilde{x}_{11} \\ \tilde{y}_{11} \end{bmatrix} = \begin{bmatrix} \tilde{a} \end{bmatrix}^{-1} * \left\{ \begin{bmatrix} I \\ J \\ K \\ L \end{bmatrix} - \begin{bmatrix} \tilde{b} \end{bmatrix} * \begin{bmatrix} \tilde{x}_2 \\ \tilde{y}_2 \\ \tilde{x}_{10} \\ \tilde{y}_{10} \end{bmatrix} - \begin{bmatrix} \tilde{c} \end{bmatrix} * \begin{bmatrix} \tilde{x}_1 \\ \tilde{y}_1 \\ \tilde{x}_9 \\ \tilde{y}_9 \end{bmatrix} \right\}
$$

where 
$$
\tilde{c} = \begin{bmatrix} b_3 & -z_3 & b_{11} & -z_{11} \\ z_3 & b_3 & z_{11} & b_{11} \\ b_{11} & -z_{11} & b_3 & -z_3 \\ z_{11} & b_{11} & z_3 & b_3 \end{bmatrix}
\qquad (9.17)
$$

$$
\begin{bmatrix} \tilde{x}_4 \\ \tilde{y}_4 \\ \tilde{x}_{12} \\ \tilde{y}_{12} \end{bmatrix} = \begin{bmatrix} \tilde{a} \end{bmatrix}^{-1} * \left\{ \begin{bmatrix} M \\ N \\ O \\ P \end{bmatrix} - \begin{bmatrix} \tilde{b} \end{bmatrix} * \begin{bmatrix} \tilde{x}_3 \\ \tilde{y}_3 \\ \tilde{x}_{11} \\ \tilde{y}_{11} \end{bmatrix} - \begin{bmatrix} \tilde{c} \end{bmatrix} * \begin{bmatrix} \tilde{x}_2 \\ \tilde{y}_2 \\ \tilde{x}_{10} \\ \tilde{y}_{10} \end{bmatrix} - \begin{bmatrix} \tilde{d} \end{bmatrix} * \begin{bmatrix} \tilde{x}_1 \\ \tilde{y}_1 \\ \tilde{x}_9 \\ \tilde{y}_9 \end{bmatrix} \right\}
$$

where 
$$
\tilde{d} = \begin{bmatrix} b_4 & -z_4 & b_{12} & -z_{12} \\ z_4 & b_4 & z_{12} & b_{12} \\ b_{12} & -z_{12} & b_4 & -z_4 \\ z_{12} & b_{12} & z_4 & b_4 \end{bmatrix}
\qquad (9.18)
$$

$$
\begin{bmatrix} \tilde{x}_5 \\ \tilde{y}_5 \\ \tilde{x}_{13} \\ \tilde{y}_{13} \end{bmatrix} = \begin{bmatrix} \tilde{a} \end{bmatrix}^{-1} * \left\{ \begin{bmatrix} \bar{A} \\ \bar{B} \\ \bar{C} \\ \bar{D} \end{bmatrix} - \begin{bmatrix} \tilde{b} \end{bmatrix} * \begin{bmatrix} \tilde{x}_4 \\ \tilde{y}_4 \\ \tilde{x}_{12} \\ \tilde{y}_{12} \end{bmatrix} - \begin{bmatrix} \tilde{c} \end{bmatrix} * \begin{bmatrix} \tilde{x}_3 \\ \tilde{y}_3 \\ \tilde{x}_{11} \\ \tilde{y}_{11} \end{bmatrix} - \begin{bmatrix} \tilde{d} \end{bmatrix} * \begin{bmatrix} \tilde{x}_2 \\ \tilde{y}_2 \\ \tilde{x}_{10} \\ \tilde{y}_{10} \end{bmatrix} - \right.
$$

$$
\left. \begin{bmatrix} \tilde{e} \end{bmatrix} * \begin{bmatrix} \tilde{x}_1 \\ \tilde{y}_1 \\ \tilde{x}_9 \\ \tilde{y}_9 \end{bmatrix} \right\} \text{ where } \tilde{e} = \begin{bmatrix} b_5 & -z_5 & b_{13} & -z_{13} \\ z_5 & b_5 & z_{13} & b_{13} \\ b_{13} & -z_{13} & b_5 & -z_5 \\ z_{13} & b_{13} & z_5 & b_5 \end{bmatrix}
\qquad (9.19)
$$

$$
\begin{bmatrix} \tilde{x}_6 \\ \tilde{y}_6 \\ \tilde{x}_{14} \\ \tilde{y}_{14} \end{bmatrix} = [\tilde{\mathbf{a}}]^{-1} * \left\{ \begin{bmatrix} \overline{E} \\ \overline{F} \\ \overline{G} \\ \overline{H} \end{bmatrix} - [\tilde{\mathbf{b}}] * \begin{bmatrix} \tilde{x}_5 \\ \tilde{y}_5 \\ \tilde{x}_{13} \\ \tilde{y}_{13} \end{bmatrix} - [\tilde{\mathbf{c}}] * \begin{bmatrix} \tilde{x}_4 \\ \tilde{y}_4 \\ \tilde{x}_{12} \\ \tilde{y}_{12} \end{bmatrix} - [\tilde{\mathbf{d}}] * \begin{bmatrix} \tilde{x}_3 \\ \tilde{y}_3 \\ \tilde{x}_{11} \\ \tilde{y}_{11} \end{bmatrix} -
$$

$$
[\tilde{\mathbf{e}}] * \begin{bmatrix} \tilde{x}_2 \\ \tilde{y}_2 \\ \tilde{x}_{10} \\ \tilde{y}_{10} \end{bmatrix} - [\tilde{\mathbf{f}}] * \begin{bmatrix} \tilde{x}_1 \\ \tilde{y}_1 \\ \tilde{x}_9 \\ \tilde{y}_9 \end{bmatrix} \right\} \text{ where } \tilde{\mathbf{f}} = \begin{bmatrix} b_6 & -z_6 & b_{14} & -z_{14} \\ z_6 & b_6 & z_{14} & b_{14} \\ b_{14} & -z_{14} & b_6 & -z_6 \\ z_{14} & b_{14} & z_6 & b_6 \end{bmatrix} \quad (9.20)
$$

$$
\begin{bmatrix} \tilde{x}_7 \\ \tilde{y}_7 \\ \tilde{x}_{15} \\ \tilde{y}_{15} \end{bmatrix} = [\tilde{\mathbf{a}}]^{-1} * \left\{ \begin{bmatrix} \overline{I} \\ \overline{J} \\ \overline{K} \\ \overline{L} \end{bmatrix} - [\tilde{\mathbf{b}}] * \begin{bmatrix} \tilde{x}_6 \\ \tilde{y}_6 \\ \tilde{x}_{14} \\ \tilde{y}_{14} \end{bmatrix} - [\tilde{\mathbf{c}}] * \begin{bmatrix} \tilde{x}_5 \\ \tilde{y}_5 \\ \tilde{x}_{13} \\ \tilde{y}_{13} \end{bmatrix} - [\tilde{\mathbf{d}}] * \begin{bmatrix} \tilde{x}_4 \\ \tilde{y}_4 \\ \tilde{x}_{12} \\ \tilde{y}_{12} \end{bmatrix} -
$$

$$
[\tilde{\mathbf{e}}] * \begin{bmatrix} \tilde{x}_3 \\ \tilde{y}_3 \\ \tilde{x}_{11} \\ \tilde{y}_{11} \end{bmatrix} - [\tilde{\mathbf{f}}] * \begin{bmatrix} \tilde{x}_2 \\ \tilde{y}_2 \\ \tilde{x}_{10} \\ \tilde{y}_{10} \end{bmatrix} - [\tilde{\mathbf{g}}] * \begin{bmatrix} \tilde{x}_1 \\ \tilde{y}_1 \\ \tilde{x}_9 \\ \tilde{y}_9 \end{bmatrix} \right\}
$$

$$
\text{where } \tilde{\mathbf{g}} = \begin{bmatrix} b_7 & -z_7 & b_{15} & -z_{15} \\ z_7 & b_7 & z_{15} & b_{15} \\ b_{15} & -z_{15} & b_7 & -z_7 \\ z_{15} & b_{15} & z_7 & b_7 \end{bmatrix} \quad (9.21)
$$

$$\begin{bmatrix}\tilde{x}_8\\ \tilde{y}_8\\ \tilde{x}_{16}\\ \tilde{y}_{16}\end{bmatrix}=[\tilde{\mathbf{a}}]^{-1}*\left\{\begin{bmatrix}\bar{M}\\ \bar{N}\\ \bar{O}\\ \bar{P}\end{bmatrix}-[\tilde{\mathbf{b}}]*\begin{bmatrix}\tilde{x}_7\\ \tilde{y}_7\\ \tilde{x}_{15}\\ \tilde{y}_{15}\end{bmatrix}-[\tilde{\mathbf{c}}]*\begin{bmatrix}\tilde{x}_6\\ \tilde{y}_6\\ \tilde{x}_{14}\\ \tilde{y}_{14}\end{bmatrix}-[\tilde{\mathbf{d}}]*\begin{bmatrix}\tilde{x}_5\\ \tilde{y}_5\\ \tilde{x}_{13}\\ \tilde{y}_{13}\end{bmatrix}-\right.$$

$$\left.[\tilde{\mathbf{e}}]*\begin{bmatrix}\tilde{x}_4\\ \tilde{y}_4\\ \tilde{x}_{12}\\ \tilde{y}_{12}\end{bmatrix}-[\tilde{\mathbf{f}}]*\begin{bmatrix}\tilde{x}_3\\ \tilde{y}_3\\ \tilde{x}_{11}\\ \tilde{y}_{11}\end{bmatrix}-[\tilde{\mathbf{g}}]*\begin{bmatrix}\tilde{x}_2\\ \tilde{y}_2\\ \tilde{x}_{10}\\ \tilde{y}_{10}\end{bmatrix}-[\tilde{\mathbf{h}}]*\begin{bmatrix}\tilde{x}_1\\ \tilde{y}_1\\ \tilde{x}_9\\ \tilde{y}_9\end{bmatrix}\right\}$$

$$\text{where }\tilde{\mathbf{h}}=\begin{bmatrix}h_8 & -z_8 & h_{16} & -z_{16}\\ z_8 & h_8 & z_{16} & h_{16}\\ h_{16} & -z_{16} & h_8 & -z_8\\ z_{16} & h_{16} & z_8 & h_8\end{bmatrix} \tag{9.22}$$

Finally, at the output of the combiner, $\tilde{\mathbf{s}}_1=[\tilde{x}_1+j\tilde{y}_1 \quad \tilde{x}_2+j\tilde{y}_2 \square \tilde{x}_8+j\tilde{y}_8]$ and $\tilde{\mathbf{s}}_2=[\tilde{x}_9+j\tilde{y}_9 \quad \tilde{x}_{10}+j\tilde{y}_{10}\square\tilde{x}_{16}+j\tilde{y}_{16}]$ are calculated. In the multiantenna configuration, the symbol vectors $\tilde{\mathbf{s}}=[\tilde{\mathbf{s}}_1 \quad \tilde{\mathbf{s}}_2 \square \tilde{\mathbf{s}}_{n_T}]$ can be obtained accurately, only if the channel parameters ($\mathbf{h}^{11} \ldots \mathbf{h}^{1n_T}, \mathbf{h}^{21} \ldots \mathbf{h}^{2n_T}, \mathbf{h}^{n_T1} \ldots \mathbf{h}^{n_Rn_T}$) are accurately estimated.

### 9.4.4 Simulation of TD-TTCM

For observing the performance of TD-TTCM, COST 207 (statistical channel models) based on the WSSUS channel models are used. TU, (non-hilly), BU (hilly), and HT are specific models of COST 207. Eight channels of TU, BU, and HT are implemented for each antenna. In the simulation, one frame size $N = 1024$ is employed. For channel estimation, eight channel coefficients of COST 207 (TU, BU, and HT) are used by GA as shown in Table 9.1 for one path delay ($\tau/T = 0.25$). It is assumed that the channel coefficients are quasi (stationary) during transmitting of one frame size ($N = 1024$ bits).

In the encoder structure as shown in Figure 9.13, two information inputs, each with 1024-bit frame size are used. To create parity bits, both encoders are constituted of three memories. 8PSK mappers are used and Transmit Diversity block has been located after the puncturer. For channel estimation, a GA is applied. For each frame $N = 1024$, 16 symbols are used as the pilot symbols for channel estimation and 1,008 symbols are used as data symbols.

**Table 9.1   Channel Coefficients, h($/T=0.25$, k) of TU, BU, and HT Types of COST 207, Which Are to Be Estimated by GA.**

| Path No. | TU | BU | HT |
|---|---|---|---|
| **(a) One Antenna** | | | |
| 1 | −1.1385 − 1.3304i | −0.3820 + 0.7018i | −0.1687 + 0.1224i |
| 2 | −1.0531 − 1.1836i | −0.6170 + 1.0973i | 0.4343 + 0.5659i |
| 3 | −0.5349 − 0.7041i | −0.6628 + 1.3421i | 0.6439 + 0.5971i |
| 4 | 0.1308 − 0.0845i | −0.4692 + 1.3192i | 0.2797 + 0.4023i |
| 5 | 0.6159 + 0.4369i | −0.1215 + 1.0312i | −0.5245 + 0.2770i |
| 6 | 0.7219 + 0.6931i | 0.1891 + 0.5906i | −1.3971 + 0.4386i |
| 7 | 0.4709 + 0.6674i | 0.2587 + 0.1598i | −1.9248 + 0.8867i |
| 8 | 0.0718 + 0.4944i | −0.0173 − 0.1242i | −1.8591 + 1.3899i |
| **(b) Two Antennas** | | | |
| 1 | 0.2506 − 0.1776i | 0.8618 − 0.6644i | −0.2453 − 0.1413i |
| 2 | 0.2096 − 0.3127i | 0.7762 − 0.8927i | −0.3950 − 0.4595i |
| 3 | 0.1451 − 0.2298i | 0.3329 − 1.0500i | −0.5242 − 0.8422i |
| 4 | 0.0190 + 0.0061i | −0.2922 − 1.0303i | −0.6263 − 1.1147i |
| 5 | −0.1572 + 0.1926i | −0.8058 − 0.8180i | −0.6753 − 1.1233i |
| 6 | −0.3249 + 0.1169i | −0.9543 − 0.4933i | −0.6361 − 0.8199i |
| 7 | −0.4058 − 0.2953i | −0.6664 − 0.1878i | −0.4903 − 0.2970i |
| 8 | −0.3412 − 0.9009i | −0.0989 − 0.0162i | −0.2605 + 0.2499i |

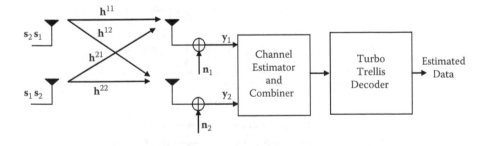

**Figure 9.13   Block diagram of decoder of TD-TTCM for 2Tx-2Rx antenna configuration [7].**

At the decoder, the signals reach the receiver antenna from the two transmitter antennas using a multipath fading environment. Quadrature and inphase coordinates of the received noisy signals are detected by using a demodulator. Then they are processed with a genetic-based channel estimator and combiner blocks. The TD-TTCM decoder as shown in Figure 9.8 evaluates the preprocessed noisy signals.

In Figures 9.14, 9.15, and 9.16, a TD-TTCM system is simulated for 8PSK over AWGN and COST 207 (TU, BU, and HT) channels. The curves are obtained for $N = 1024$ and different SNR and iteration values with antenna configurations of 2Tx-1Rx and 2TX-2Rx.

For a given SNR value, as the iteration number increases, performance gets better. The performance results of Figure 9.14 through Figure 9.16 are compared for $2T_x$-$1R_x$ and $2T_x$-$2R_x$ configurations. In the simulations, for $2T_x$-$2R_x$ configurations, at 6 dB, the best BER of $10^{-5}$ is obtained for TU in Figure 9.14. In Figure 9.15 and Figure 9.16, the performances of $2T_x$-$2R_x$ are about BER of $10^{-5}$ at 7 dB for BU and BER of $10^{-5}$ at 10 dB for HT, respectively. In all figures, the performance of $2T_x$-$2R_x$ is superior to the performance of $2T_x$-$1R_x$ at all SNR values. In HT type COST 207 channels, the performance of $2T_x$-$2R_x$ (3rd iteration) reaches BER of $10^{-4}$ at 9.5 dB, while $2T_x$-$1R_x$ (3rd iteration) reaches the same BER at 12 dB, thus coding gain is about 2.5 dB (Figure 9.16). In TU and BU types of COST 207

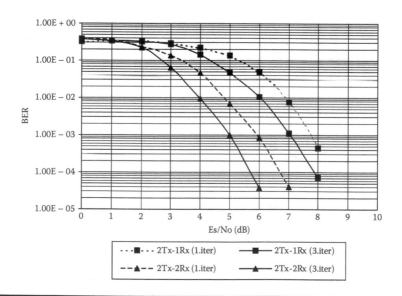

**Figure 9.14   TD-TTCM for TU type of COST 207 [7].**

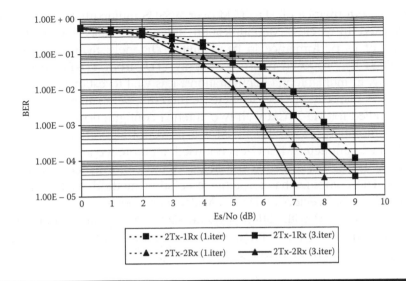

**Figure 9.15    TD-TTCM for BU type of COST 207 [7].**

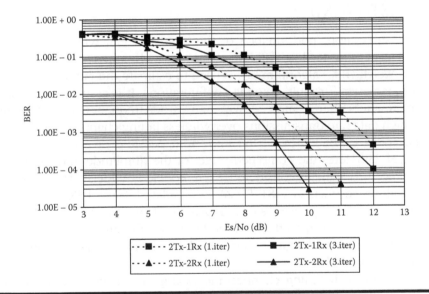

**Figure 9.16    TD-TTCM for HT type of COST 207 [7].**

channels, coding gains above 2 dB are observed at BER of $10^{-4}$ in Figure 9.14 and Figure 9.15, respectively. The performance of TD-TTCM over HT is worse than TU and BU due to multipath fading environments.

Here, performance of TD-TTCM over WSSUS multipath MIMO channels is evaluated. In wireless communication, antenna diversity is very attractive because of its performance in band-limited channels. Symbol vectors are transmitted by proposed transmit diversity (TD) from TD-TTCM encoder. Especially, in multiantenna configurations, it is vital to estimate the channel parameters correctly in order to increase the error performance. A modified GA is studied for TD-TTCM with two antenna configurations. It is observed that our modified GA for TD-TTCM is a powerful algorithm for estimating the channel parameters. If a GA is not used, interference and noise become dominant in BER performance and degrades the performance in all channel models. The combiner has an iterative solution method, and it offers the estimated symbol vectors to TD-TTCM decoder. The performance of TTCM signals is simulated with different frame size $N$ and iteration numbers. In all simulations, a $2T_x$-$2R_x$ configuration offers up to 2.5 dB significant additional gain in comparison to a $2T_x$-$1R_x$ configuration.

## 9.5 Space Time Turbo Trellis-Coded Modulation (ST-TTCM)

In this subsection, performance of the ST-TTCM [35] will be evaluated over Rician and Rayleigh fading channels with imperfect phase. The Baum–Welch (BW) algorithm [23] is used to estimate the fading and phase jitter parameters for multi-antenna configurations. It is assumed that channel parameters change more slowly than carrier frequency. It is clear that with high data rate transmissions over wireless fading channels, space–time block codes (STBC) provide the maximum possible diversity advantage [21,22,24,25]. Here, the combined effect of the amplitude and the phase of the received signal is considered, each modeled by Rician and Tikhonov distributions respectively. ST-TTCM is simulated for 8-PSK for several Rician factor $K$ and phase distortion factor $\eta$. Thus, the results reflect the degradations due to the effects of the fading on the amplitude and phase noise of the received signal while the channel parameters are estimated by the BW algorithm.

In a multipath fading channel, increasing the bit error performance or reducing the effective error rate is extremely difficult. In an AWGN environment, using typical modulation and coding schemes, reducing the effective bit error rate from $10^{-2}$ to $10^{-3}$ may require only 1 or 2 dB higher SNR. Achieving the same in a multipath fading environment may require an extra SNR up to 10 dB [3–5,18–20]. The preferred method to improve SNR may not be increasing transmitting power or bandwidth, which is contrary to the requirements of next-generation systems. Therefore it is crucial to effectively reduce the effect of fading at both the remote units and the base stations, without additional power or any sacrifice in bandwidth.

Recently, different transmit diversity techniques have been introduced to benefit from antenna diversity also in the downlink, while putting the diversity burden on the base station. In [17], space–time trellis coding was introduced proposing joint design of coding, modulation, and transmitter and receiver diversity. Furthermore in [21], TTCM and STBC are concatenated and in [22] multi-user space–time schemes are introduced. Substantial benefits can be achieved by channel coding techniques appropriate to multiple transmit antenna systems.

To improve bit-error performance and bandwidth efficiency of communication systems, estimation of unknown linear and nonlinear parameters, such as channel parameters, become important [26–28]. The BW algorithm was developed for and is widely employed to adjust the model parameters of hidden Markov model's (HMM) for finding the maximum likelihood (ML) estimation of the distribution parameters of a Markov chain probabilistic function [23].

Here, performance of ST-TTCM is evaluated taking fading and phase jitter into account. The general block diagram of the considered scheme is given in Figure 9.17. The input binary data feeds a turbo encoder and symbol stream **x**, passes through a space–time block encoder.

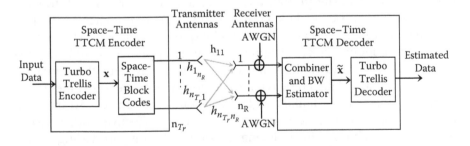

**Figure 9.17  Block diagram of ST-TTCM (Copyright permission by John Wiley & Sons Ltd. [35]).**

## 9.5.1 ST-TTCM Encoder

In the encoder of ST-TTCM, the input sequence of information bit pairs is encoded in an Ungerboeck-style encoder to yield the M-PSK sequence. The information bits are interleaved on a pairwise basis and encoded again into the sequence. The second encoder's output symbols are deinterleaved to ensure that the ordering of the two information bits partly defining each symbol corresponds to that of the first encoder. Finally, the first symbol of the first encoder, the second symbol of the second encoder, the third symbol of the first encoder, and the fourth symbol of the second encoder are transmitted, and this process continues in this order. By using an interleaver and deinterleaver for the second encoder, each symbol index $k$ before the selector has the property of being associated with input information bit group index $k$. The interleaver must map even positions to even positions and odd positions to odd ones. This ensures that each information bit influences either the state of the upper or lower encoder but never both.

At the receiver side, the distorted multipath signals arrive to $n_R$ number of receiver antennas. Here, channel parameters are estimated by a modified Baum–Welch algorithm and hence combiner output $\tilde{\mathbf{x}}$ feeds the turbo-trellis decoder. And finally, the input data sequence is estimated.

In wireless communication, multipath fading increases both severe amplitude and phase distortion. So, it is crucial to combat the effect of the multipath fading at both the remote units and the base stations, without additional power or any sacrifice in bandwidth. Because the transmitter has no information about the channel fading characteristics, a feedback connection can be installed between receiver and transmitter that causes complexity in both transmitter and receiver systems. In some cases, there may be no link between them. To overcome this lack of information, there are some known techniques such as spread spectrum, time, and frequency diversity. Spread-spectrum approaches are ineffective if the coherent bandwidth of the channel is larger than the spreading bandwidth. Using time interleaving separately may cause long time delays. Frequency diversity is limited with bandwidth occupancies. So in fading environments, space–time coding is a practical, effective and, hence, a widely used technique for reducing the effect of multipath fading [24,25]. Space–time coding is a bandwidth and power-efficient method of communication over Rician fading channels that achieve the benefits of a multiple transmit antenna.

A space–time block encoder maps the input symbols on entries of a $p \times n_{Tr}$ matrix $\mathbf{G}$, where $n_{Tr}$ is the number of transmit antennas. The entries of the matrix $\mathbf{G}$ are the Tr complex symbol $x_i$, their complex conjugate $x_i^*$, and linear combinations of $x_i$ and $x_i^*$. The $p \times n_{Tr}$ matrix $\mathbf{G}$, which defines the space–time block code, is a complex generalized orthogonal as defined in [17]. This means that the columns of **G** are orthogonal. The imperfect phase reference effect of the fading channel is modeled as

$$b = \rho \, e^{j\theta} \qquad (9.23)$$

$\rho$ is fading amplitude and the term $e^{j\theta}$ is a unit vector where $\theta$ represents the phase noise, mentioned in [26–28], which is assumed to have Tikhonov pdf given by

$$p(\theta) = \frac{e^{\eta\cos(\theta)}}{2\pi I_0(\eta)} \qquad |\theta| \le \pi \qquad (9.24)$$

where $I_0$ is the modified Bessel function of the first-order zero and $\eta$ is effective signal-to-noise ratio of the phase noise. $\rho$ has Rician distribution which can be written as [29–32]

$$P(\rho) = 2\rho(1+K) \; e^{(-\rho^2(1+K)-K)} I_0\left[\rho 2 \sqrt{K(1+K)}\right] \qquad (9.25)$$

$K$ is the SNR of the fading effect of the channel. If the $K$ factor is infinity, the channel is only AWGN and there is no fading effect; if $K$ is zero, it means that the channel is in the worst case and this type of channel is called the Rayleigh channel. In a multi-antenna configuration, the symbols $\tilde{\mathbf{x}} = \{\tilde{x}_1, \tilde{x}_2, \tilde{x}_3, ..., \tilde{x}_{n_T}\}$ at the output of the combiner can be obtained accurately, only if the channel parameters $(b_{1,1} ... b_{1,n_R}, b_{2,1} ... b_{2,n_R} ... ... b_{n_T,1} ... b_{n_T,n_R})$ are correctly estimated. Thus in the following section, the modified Baum–Welch algorithm—based on blind equalization techniques—is provided to estimate these parameters for an ST-TTCM scheme.

## 9.5.2 Blind Equalization

Consider the block diagram of the ST-TTCM scheme shown in Figure 9.17. At the $i$th epoch, $m$-bit information vector $\mathbf{d}_i = (d_i^1, d_i^2, ... d_i^m)^T$ is fed as input to TTCM system, followed by $m/(m+1)$ convolutional encoder and M-PSK mappers. The output of this system is denoted as $g(\mathbf{d}_i)$. The received ST-TTCM coded signals over a Rician fading channel with imperfect phase, at the signaling instant has the form

$$y_i = \rho_i g(\mathbf{d}_i) e^{j\theta_i} + n_i \qquad (9.26)$$

where $x(i) = g(\mathbf{d}_i)$.

The following assumptions are imposed on a received signal model:

**AS1:** $n_i$ is a zero mean, i.i.d., additive Gaussian noise sequence with variance.

**AS2:** $\mathbf{d_i}$ is an $m$ digit binary, equiprobable input information sequence to the TTCM.

**AS3:** $\rho_i$ and $\theta_i$ represent a normalized fading and phase parameters having Rician and Tikhonov probability density functions for the signaling interval.

Based on assumption **AS1,** the maximum likelihood metric for $\mathbf{d}$ is proportional to the conditional probability density function of the received data sequence of length $N$ (given $\mathbf{d}$) [33] and is of the form

$$m_s(y,d) = \sum_{i=1}^{N} \left| y_i - \rho_s g(\mathbf{d_i}) e^{j\theta_s} \right|^2 \tag{9.27}$$

where $\mathbf{D} = (\mathbf{d_1}, \mathbf{d_2}, \dots, \mathbf{d_N})$. Because the transmitted data sequence is not available at the receiver, the ML metric to be maximized can be obtained after averaging $m_s(y,d)$ over all possible transmitted sequences $\mathbf{d}$,

$$m_s(y) = \sum_{d} m_s(y,d).$$

Then the ML estimator of $\quad = [\rho, \theta, \sigma^2_n]$ is the one that maximizes average ML metric $\hat{\quad} = \arg \max[m_s(y)]$. Here slowly varying fading channels are assumed; $\rho$ and $\theta$ remain constant over the observation interval. The metric for $\mathbf{D}$ can be obtained by averaging Equation (9.27) over all possible values of $\mathbf{d_i}$,

$$m_s(y) = \sum_{d} \sum_{i=1}^{N} \left| y_i - \rho_s g(\mathbf{d_i}) e^{j\theta_s} \right|^2 \tag{9.28}$$

Fortunately, a general iterative approach to compute ML estimates known as the Baum–Welch (BW) algorithm [23] can be used to significantly reduce the computational burden of an exhaustive search.

Because there are $n_T$ signals that are superposed in the receiver antennas, the auxiliary function of BW algorithm can be modified for space–time block codes as follows:

$$\Theta(\hat{}^{(t)}, \text{''}) = C + \sum_{i=1}^{N}\sum_{m=1}^{P}\gamma_{i,m}^{(t)}\left\{\frac{1}{\sigma_v'^2}\left|r_i - \rho_s'e^{j\theta_s'}(\sum_{u=1}^{n_T}g_u(\phi_m))\right|^2\right\} \quad (9.29)$$

where $\gamma_{i,m}^{(t)} = f_{\phi'^{(t)}}(r, \phi_i = \phi_k)$ is the joint likelihood of the data $r$ and being at state $\phi_k$ at time $i$. $g_1(.)$ through $g_{n_T}(.)$ are the functions that give the signal outputs from the antennas at state $\phi_m$. Here it is assumed that all the signals follow the same path with the distortion of $\rho_s'e^{j\theta_s'}$.

Using the definitions of forward $\alpha_{i,k}$ and backward $\beta_{i,k}$ variables, which are the probabilities of the partial observation sequences, given the state $\phi_k$ at time $i$, $\gamma_{i,k}$ can be recursively obtained as [9,23,31,34]

$$\gamma_{i,k}^{(t)} = \alpha_{i,k}^{(t)}\beta_{i,k}^{(t)} \quad (9.30)$$

The iterative estimation formulas given below can be derived from the derivation of the auxiliary function in Equation (9.29). Thus, channel parameters are estimated by setting the gradient functions to zero $\nabla_{\rho_i'}\Theta = 0$, $\nabla_{\theta_i'}\Theta = 0$, $\nabla_{\sigma_v'^2}\Theta = 0$.

$$\nabla_{\rho_i'}\Theta = \frac{2}{\sigma_v'^2}\sum_{i=1}^{N}\sum_{m=1}^{P}\gamma_{i,m}^{(t)}$$

$$\cdot \ e\left\{\left[y_i - \rho_s'e^{j\theta_s'}(\sum_{u=1}^{n_T}g_u(\phi_m))\right](\sum_{u=1}^{n_T}g_u(\phi_m))^*\right\} \quad (9.31)$$

$$\nabla_{\theta_i'}\Theta = \frac{2}{\sigma_v'^2}\sum_{i=1}^{N}\sum_{m=1}^{P}\gamma_{i,m}^{(t)}$$

$$\cdot \ e\left\{\left[y_i - \rho_s'e^{j\theta_s'}(\sum_{u=1}^{n_T}g_u(\phi_m))\right]je^{-j\theta_s'}\rho_s'(\sum_{u=1}^{n_T}g_u(\phi_m))^*\right\} \quad (9.32)$$

$$\nabla_{\sigma_v'^2}\Theta = \frac{N}{2\sigma_v'^2} + \frac{N}{2\sigma_v'^4}\sum_{i=1}^{N}\sum_{m=1}^{P}\gamma_{i,m}^{(t)} \ e\left\{\left[y_i - \rho_s'e^{j\theta_s'}(\sum_{u=1}^{n_T}g_u(\phi_m))\right]^2\right\} \quad (9.33)$$

There is no limit to the number of transmitter and receiver antennas. At each receiver antenna of $n_R$ is the number of receiver antennas, the sum of the signals is received, which is transmitted from $n_T$ number of antennas at the same instant.

Here, space–time block codes are represented for turbo trellis-coded signals, while a modified blind algorithm estimates the channel parameters; $\sigma$, $\rho$, and $\theta$ by using Equations (9.29)–(9.33). The results will reflect the degradations due to the effects of the fading on the amplitude of the received signal as well as phase jitter.

### 9.5.3 Simulation Results of ST-TTCM for Two-Transmitter and One-Receiver Antenna Configuration

In this subsection, to emphasize the performance of ST-TTCM over fading channels with imperfect phase, the diversity scheme is simplified as two transmitter antennas and one receiver antenna instead of multi-antenna configurations. At the transmitter side, 8-PSK modulation is chosen as in Figure 9.18.

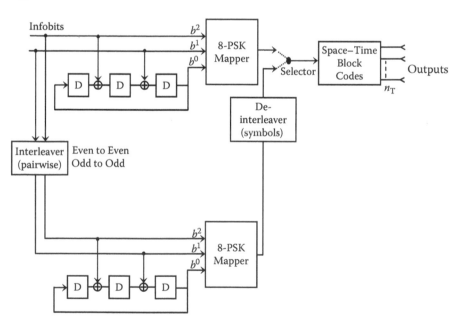

**Figure 9.18** **Encoder of ST-TTCM for 8-PSK. (Copyright permission by John Wiley & Sons Ltd. [35].)**

In the encoder structure, there are two information inputs, each with 1024 bit frame size. To create parity bits, both encoders are constituted of three memories. 8-PSK mappers are used, and space-time block has been located after the punctures.

In this model there are only two channels and thus, we use $h_1$ and $h_2$ terms for $h_{1,1}$ and $h_{2,1}$, respectively, as shown in Figure 9.19. The multipath channels $h_1$ and $h_2$ are modeled as,

$$h_1(t) = h_1(t + T) = \rho_1 e^{j\theta_1}$$
$$h_2(t) = h_2(t + T) = \rho_2 e^{j\theta_2} \tag{9.34}$$

where $T$ is the symbol duration. The outputs of combiner for the two antennas [17], given below, are now the entries for our turbo trellis decoder (Figure 9.19).

$$\tilde{x}_1 = (\rho_1^2 + \rho_2^2)x_1 + h_1^* n_1 + h_2 n_2^*$$
$$\tilde{x}_2 = (\rho_1^2 + \rho_2^2)x_2 - h_1 n_2^* + h_2^* n_1 \tag{9.35}$$

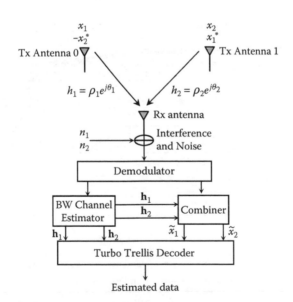

**Figure 9.19  Block diagram of decoder of ST-TTCM for 2Tx-1Rx antenna configuration. (Copyright permission from John Wiley & Sons Ltd. [35].)**

There are two transmitter antennas and one receiver antenna. In other words, the antenna structure is 2Tx-1Rx and these two paths are assumed to have the same properties. Received signals pass through the Baum–Welch channel estimator and combiner. Here, it is assumed that the variation of the channel parameters is ten times slower than the carrier frequency and 15 iterations are assumed to be enough to find the channel parameters by the BW algorithm. The preprocessed noisy signals are evaluated by the turbo trellis decoder as shown in Figure 9.19. To emphasize the importance of imperfect-phase and fading effects, the performance of the considered scheme is simulated for various η and $K$ values

As an example, a ST-TTCM system is simulated for 8-PSK over a AWGN and Rician channel with phase jitter. Various curves are obtained for different iteration numbers, $K$, η and SNR values. Figure 9.20 shows the performance of ST-TTCM signals for $K$ = ∞ dB with η = 10, 20 and ∞ dB. Here, dashed lines show the performance when perfect channel estimation is assumed and others show the performance when channel parameters are estimated by BW algorithm given in Equations 9.31–9.35. It can be easily seen that the BER results of the system are very close to that of the perfect-channel-estimation case when the BW algorithm is used. Since

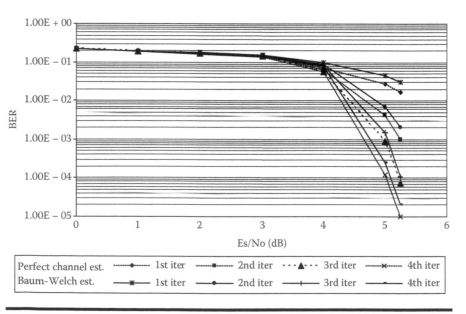

**Figure 9.20   AWGN channel, 2Tx-1Rx, η = ∞, 20, 10 dB, perfect and Baum–Welch channel parameter estimations. (Copyright permission by John Wiley & Sons Ltd. [35].)**

the channel parameters are well estimated by the BW algorithm, there is not any degradation on the BER when η alters. Because phase jitter, $e^{j\theta}$, only affects the noise as shown in Equation (9.43), and does not change the noise power, it only rotates the noise phase according to θ. This property can be seen for other $K$ values as in Figures 9.21 and 9.22.

In Figure 9.21, BER performances are evaluated with no channel state information (CSI).

Because there is no BW estimator, BER changes as η factor gets different values. The BER performance is drastically affected as η changes 20 dB to 10 dB after 6 dB Es/No values. Figure 9.22 shows the performance of ST-TTCM signals for $K = 10$ dB with η = 10, 20, and ∞ dB. Here, dashed lines show the performance when perfect channel estimation is assumed and others show the performance when channel parameters are estimated by BW algorithm. The error performance is worse than the AWGN case because both ρ and θ will be estimated.

In Figure 9.23, BER performances are evaluated in no CSI. Because there is no BW estimator, BER results become worse as η decreases. Similar results are obtained for Rayleigh channel as in Figures 9.24 and 9.25. The simulation curves in Figure 9.24 confirm Figure 9.22 for the first two iterations, but better BER values are expected after further iterations.

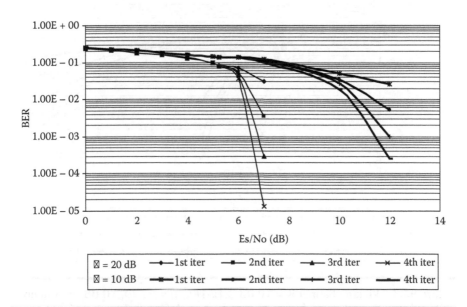

**Figure 9.21** AWGN channel, 2Tx-1Rx, η = 20, 10 dB, no channel state information. (Copyright permission by John Wiley & Sons Ltd. [35].)

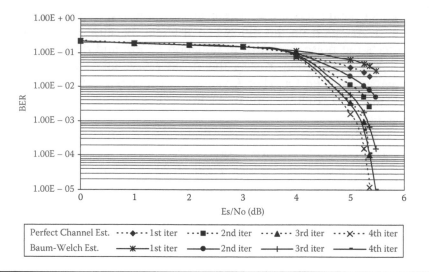

**Figure 9.22** **Rician channel K = 10 dB, 2Tx-1Rx, $\eta = \infty$, 20, 10 dB, perfect, and Baum–Welch channel parameter estimation. (Copyright permission by John Wiley & Sons Ltd. [35].)**

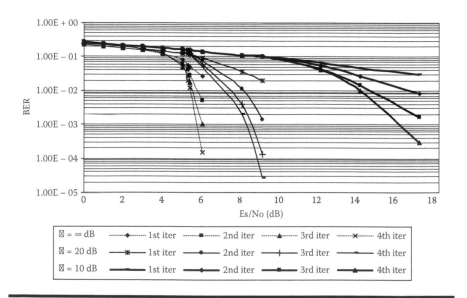

**Figure 9.23** **Rician channel K = 10 dB, 2Tx-1Rx, $\eta = \infty$, 20, 10 dB, no channel state information. (Copyright permission by John Wiley & Sons Ltd. [35].)**

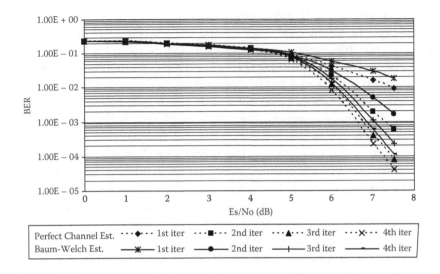

**Figure 9.24** **Rayleigh channel, 2Tx-1Rx, η = ∞, 20, 10 dB, perfect and Baum–Welch channel parameter estimations. (Copyright permission by John Wiley & Sons Ltd. [35].)**

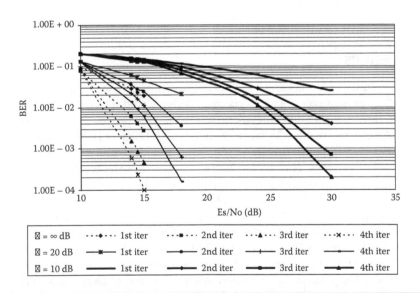

**Figure 9.25** **Rayleigh channel, 2Tx-1Rx, η = °, 20, 10 dB, no channel state information. (Copyright permission by John Wiley & Sons Ltd. [35].)**

The importance of the channel estimation, especially estimation of the phase noise θ, can be determined after comparing simulation results. The complexity of the BW algorithm is related to the number of channels. Each additional increment in the number of transmitter antennas results in the addition of an extra $n_R$ number of channels. In the same way, each additional increment in the number of receiver antennas results in the addition of an extra $n_T$ number of channels. The BW complexity is proportional to the antenna configuration structure, whereas the iterative decoding process is not affected by the increment in the number of channels.

In space–time block coding (STBC), each symbol is transmitted from every transmitter antenna; in other words, each symbol is sent more than once, which is related to the number of transmitter antennas. For this reason, an STBC with more antennas not only combats with the effects of fading and imperfect phase, but also reduces the error probability to the lower bound limits as given in [9]. The modified BW algorithm for ST-TTCM is capable of estimating the channel phase and fading distortions. If the BW algorithm is not used, phase distortion becomes dominant in BER performance and degrades the performance in all channel models. In different η, $K$ and SNR values, as the iteration number increases, performance improves.

# References

[1] Berrou, C., A. Glavieux, and P. Thitimasjshima. Near Shannon-Limit Error Correcting Coding and Decoding: Turbo Codes (1). *IEEE International Conference on Communication*:1064–1070, Geneva, Switzerland, May 1993.

[2] Robertson, P., and T. Worz T. Bandwidth-Efficient Turbo Trellis-Coded Modulation Using Punctured Component Codes. *IEEE J., SAC* 16(2), 1998.

[3] Hideki, O., and H. Imai H. Performance of the Deliberate Clipping with Adaptive Symbol Selection for Strictly Band-Limited OFDM Systems. *IEEE Journal Selected Areas* 18(11):2270–2277, 2000.

[4] U•an, O. N., and O. Osman O. Turbo Trellis Coded Modulation (TTCM) with Imperfect Phase Reference. *Istanbul University-Journal of Electrical and Electronics Engineering (IU-JEEE)* 1:1–5, 2001.

[5] U•an, O. N., O. Osman, and S. Paker S. Turbo Coded Signals over Wireless Local Loop Environment. *International Journal of Electronics and Communications (AEU)* 56(3):163–168, 2002.

[6] Blackert, W., and S. Wilson S. Turbo Trellis Coded Modulation. *Conference Information Signals and System CISS*, 1996.

[7] Gose, E., H. Pastaci, O. N. U•an, K. Buyukatak, and O. Osman O. Performance of Transmit Diversity-Turbo Trellis Coded Modulation (TD-TTCM) over Genetically Estimated WSSUS MIMO Channels. *Frequenz* 58:249–255, 2004.

[8]  Gose, E. Turbo Coding over Time Varying Channels. Ph.D. Thesis, Yildiz Technical University, Turkey, 2005.

[9]  Foschini, G., and M. Gans. On Limits of Wireless Communications in a Fading Environment when Using Multiple Antennas. *Wireless Personal Commun.* 6(3):311–335, Mar. 1998.

[10]  Telatar, I. E. Capacity of Multi-Antenna Gaussian Channels. *Eur. Trans. Telecom. (ETT)* 10(6):585–596, Nov. 1999.

[11]  Wittneben, A. Base Station Modulation Diversity for Digital SIMUL-CAST. in *Proc.IEEE Vehicular Technology Conf. (VTC)*: 505–511, May 1991.

[12]  Bello, P. A. Characterization of Randomly Time-Variant Linear Channels. *IEEE Trans. on Communication Systems* CS-11:360–393, Dec. 1963.

[13]  Hoeher, P. A Statistical Discrete-Time Model for the WSSUS Multi Path Channel. *IEEE Trans. on Vehicular Technology* VT-41(4):461–468, April 1992.

[14]  COST 207. *Digital Land Mobile Radio communications.* Office for Official Publications of the European Communities, Abschlussbericht, Luxemburg, 1989.

[15]  Holland, J. *Adaptation in Natural and Arti"cial Systems.* Cambridge, MA: MIT Press, 1975.

[16]  Goldberg, D. E. *Genetic Algorithms in Search, Optimization, and Machine Learning.* Reading, MA: Addison-Wesley, 1989.

[17]  Alamouti, S. M. A Simple Transmit Diversity Technique for Wireless Communication. *IEEE Journal of Selected Areas in Communication* 16(8), Oct. 1998.

[18]  Osman, O. Performance of Turbo Coding. Ph.D. Thesis, Istanbul University,Turkey, 2003.

[19]  U•an, O. N. Performance Analysis of Quadrature Partial Response-Trelllis Coded Modulation (QPR-TCM). Ph.D. Thesis, Istanbul Technical University, 1994..

[20]  Buyukatak, K. Evaluation of 2-D Images for Turbo Codes. Ph.D. Thesis, Istanbul Technical University, 2004.

[21]  Bauch, G. Concatenation of Space-Time Block Codes and Turbo-TCM. *Proc. ICC'99,* Vancouver.

[22]  Lu, B., and X. Wang X. Iterative Receivers for Multi-User Space-Time Coding Schemes. *IEEE JSAC* 18:11, 2000.

[23]  Baum, L.E., T. Petrie, G. Soules, and N. Weiss. A Maximization Technique occurring in the Statistical Analysis of Probabilistic Functions of Markov Chains. *The Annals of Mathematical Statistics* 41:164–171, 1970.

[24]  Tarokh, V., N. Seshadri, and A. Calderbank A. Space-Time Codes for High Data Rate Wireless Communication: Performance Criterion and Code Construction. *IEEE Transaction on Information Theory* 44:744–765, 1998.

[25]  Tarokh, V., H. Jafarkhani, and A. Calderbank. Space-Time Codes from Orthogonal Design. *IEEE Transaction on Information Theory,* June 1999.

[26]  U•an, O. N., O. Osman, and A. Gumus A. Performance of Turbo Coded Signals over Partial Response Fading Channels with Imperfect Phase Reference. *Istanbul University-Journal of Electrical & Electronics Engineering (IU-JEEE)* 1(2):149–168, 2001.

[27] U•an, O. N. Trellis Coded Quantization/Modulation over Mobile Satellite Channel with Imperfect Phase Reference. *International Journal of Satellite Communications* 16:169–175, 1998.

[28] Stavroulakis, P. *Wireless Local Loop: Theory and Applications.* John Wiley and Sons Ltd. Publications, 2001.

[29] Divsalar, D., and M. K. Simon MK. The Design of Trellis Coded M-PSK for Fading Channels: Performance Criteria. *IEEE Transaction on Communication* 36:1004–1012, Sept. 1998

[30] Erkurt, M, and J. G. Proakis JG. Joint Data Detection and Channel Estimation for Rapidly Fading Channels. in *IEEE Globecom*:910–914, 1992.

[31] Divsalar, D., M. K. Simon. Trellis Coded Modulation for 4800-9600 Bits/s Transmission over a Fading Mobile Satellite Channel. *IEEE J. Select. Areas in Commun.* SAC-5:162–175, May 1987.

[32] U•an, O. N., and H. A. Cirpan HA. *Blind Equalisation of Quadrature Partial Response-Trellis Coded Modulated Signals Chapter of Third Generation Mobile Telecommunications.* Springer Verlag, 2001.

[33] Rabiner, L. R. A Tutorial on Hidden Markov Models and Selected Applications in Speech Recognition. *Proceeding of the IEEE* 77:257–274, 1989.

[34] Kaleh, K., R. and Vallet. Joint Parameter Estimation and Symbol Detection for Linear or Nonlinear Unknown Channels. *IEEE Trans. on Commun.* 42:2406–2413, July 1994.

[35] U•an, O. N., and O. Osman. Concatenation of Space-Time Block Codes and Turbo Trellis Coded Modulation (ST-TTCM) over Rician Fading Channels with Imperfect Phase. *International Journal of Communication Systems* 17:347–361, 2004.

# Chapter 10

# Low-Density Parity Check Codes

In 1963, Gallager [1] invented low-density parity check (LDPC) codes, providing low encoder and decoder complexity. LDPC with constrained random code ensembles and iterative decoding algorithms are serious competition for turbo codes in terms of complexity and performance. Until 1996, Reed–Solomon (RS) and convolutional codes were considered perfectly suitable for error-control coding. There were only a few studies on LDPC codes, such as Zylabov and Pinkster [2], Margulis [3], and Tanner [4]. Then MacKay and Neal [5], and Wiberg [6] rediscovered LDPC codes. There has been a great amount of research on this subject because the decoding of LDPC codes is quicker and less complex than the decoding of turbo codes; however, the coding process is slow by the square of code length $n$. As originally suggested by Tanner [4], LDPC codes are well represented by bipartite graphs in which one set of nodes, the *bit nodes*, corresponds to elements of the code word and the other set of nodes, the *check nodes*, corresponds to the set of parity-check constraints that define the code. *Regular* LDPC codes are those for which all nodes of the same type have the same degree.

For example, Figure 10.1(a) represents (3,4) a regular LDPC code in which all bit nodes have degree 3 and all check nodes have degree 4. *Irregular* LDPC codes have variable bit and check node degrees and they were introduced in [7], [8]. As an example, in Figure 10.1(b), an irregular LDPC code structure is given, in which the degree of the first half of the bit nodes is 2 and the other half is 3.

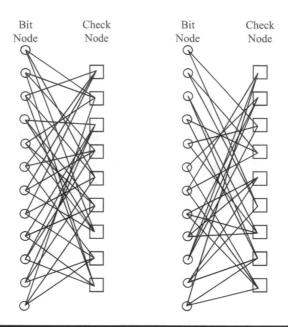

**Figure 10.1  Various bipartite representations of LDPC codes.**

In band-limited channels, such as deep space and satellite communications, continuous phase modulation has explicit advantages because it has a low-spectral occupancy property. CPM is composed of a continuous phase encoder (CPE) and a memoryless mapper (MM). The CPE is a convolutional encoder producing code word sequences that are mapped onto waveforms by the MM, creating a continuous phase signal. CPE-related schemes have better performance than systems using the traditional approach for a given number of trellis states, because they increase the Euclidean distance. Once the memory structure of CPM is assigned, it is possible to design a single-joint convolutional code, composed of trellis and convolutionally coded CPM systems as in [9,10]. To improve error performance and bandwidth efficiency, low-density parity check codes and continuous phase modulation are combined. Low-density parity check/continuous phase frequency shift keying (LDPC/CPFSK) is introduced in [11].

An LDPC code is a linear block code that is specified by a very sparse parity check matrix. An LDPC code can be represented easily by a bipartite graph while parity check matrix $H$ has $N$ columns and $M$ rows. The bipartite graph of this code has $N$ bit nodes, $M$ check nodes, and a certain number of edges. The code rate $R$ is $1-(M/N)$. Each bit node represents a bit symbol in the code words and each check node represents a parity equation of code. There is a line drawn in the bipartite graph between a bit node and a check node if, and only if, that bit is involved in that equation. LDPC codes are defined by the number of 1 bits in the parity check matrix's rows and columns. LDPC codes are divided into two groups, regular LDPC codes and irregular LDPC codes. Regular LDPC codes are those in which all rows and columns of the parity check matrix have the same degree. In an irregular LDPC code, the degrees of each set of nodes are chosen according to some distribution rule. Regular A $(j,k)$ LDPC code has a bipartite graph in which all bit nodes have degree $j$ and all check nodes have degree $k$. That means, in the parity check matrix $H$, all the row weights are $k$ and all the column weights are $j$. The partite graph of a $(3,4)$ regular LDPC code is shown in Figure 10.1 (a) and the parity check matrix of this code is given in Equation (10.1).

$$
H = \begin{bmatrix}
0 & 0 & 1 & 0 & 0 & 1 & 1 & 1 & 0 & 0 & 0 & 0 \\
1 & 1 & 0 & 0 & 1 & 0 & 0 & 0 & 0 & 0 & 0 & 1 \\
0 & 0 & 0 & 1 & 0 & 0 & 0 & 0 & 1 & 1 & 1 & 0 \\
0 & 1 & 0 & 0 & 0 & 1 & 1 & 0 & 0 & 1 & 0 & 0 \\
1 & 0 & 1 & 0 & 0 & 0 & 0 & 1 & 0 & 0 & 1 & 0 \\
0 & 0 & 0 & 1 & 1 & 0 & 0 & 0 & 1 & 0 & 0 & 1 \\
1 & 0 & 0 & 1 & 1 & 0 & 1 & 0 & 0 & 0 & 0 & 0 \\
0 & 0 & 0 & 0 & 0 & 1 & 0 & 1 & 0 & 0 & 1 & 1 \\
0 & 1 & 1 & 0 & 0 & 0 & 0 & 0 & 1 & 1 & 0 & 0
\end{bmatrix} \qquad (10.1)
$$

The optimal distribution of the rows and columns of an irregular LDPC code can be computed by density evolution. An irregular LDPC code can be specified by a degree distribution $(\lambda, \rho)$ pair or

$$
\lambda(x) = \sum_{i=2}^{d_{l_{max}}} \lambda_i x^{i-1}
$$

$$
\rho(x) = \sum_{i=2}^{d_{r_{max}}} \rho_i x^{i-1}
$$

(10.2)

generating functions. Here, $\lambda_i$ and $\rho_i$ are the fraction of edges with (left, right) degree $i$ respectively. We define the (left, right) degree of an edge as the degree of the corresponding (left, right) node it is connected to. $d_{l_{max}}$ is the maximal left degree of any edges and $d_{r_{max}}$ is the maximal right degree of any edges. The partite graph for an irregular LDPC code is shown in Figure 10.1 (b). For this code, $\lambda_2 = 0.4$, $\lambda_3 = 0.6$, $\rho_3 = 0.6$, and $\rho_4 = 0.4$.

## 10.1 Encoding of LDPC Codes

The parity check matrix (*H*) of an LDPC is randomly constructed. Therefore, the code words of the LDPC code *A* with length *n* must be written in systematic form

$$\vec{v} = \left(\bar{p} \mid \bar{u}\right) \tag{10.3}$$

where, $\bar{u}$ is the *k*-tuple message bit and $\bar{p}$ is the (*n-k*)-tuple parity-check bits. Let the parity-check matrix of *A* be *H*, can be defined as Equation (10.4), where $A_1$ and $A_2$ are (*n-k*) × (*n-k*) and *k* × (*n-k*) matrices.

$$H^T = \begin{bmatrix} A_1 \\ A_2 \end{bmatrix} \tag{10.4}$$

Since

$$\vec{v}.H^T = \left(\bar{p} \mid \bar{u}\right) \begin{bmatrix} A_1 \\ A_2 \end{bmatrix}$$

then

$$\bar{p}.A_1 + \bar{u}.A_2 = 0 \tag{10.5}$$

Let the generator matrix of *A* be *G* in the form of $(\bar{\bar{p}} \mid I_k)$, where $\bar{\bar{p}}$ is *k* × (*n-k*) matrix. Because $\vec{v} = \bar{u}.G = \bar{u}.\left[\bar{\bar{p}} \mid I_k\right]$ we have $p = u.\bar{\bar{p}}$. Using Equation (10.5) for any $\bar{u}$, we have,

$$\bar{u}\left(\bar{\bar{p}}A_1 + A_2\right) = 0 \tag{10.6}$$

Thus we have $\bar{\bar{p}} = A_2 A_1^{-1}$ and a generator matrix in symmetric form is described in Equation (10.7).

$$G = \left[\bar{\bar{p}} \mid I_k\right] = \left[A_2 A_1^{-1} \mid I_k\right] \tag{10.7}$$

The code word can be found by multiplying the message bits with the generator matrix $G(\bar{u}.G)$.

## 10.2 Decoding of LDPC Codes

The decoding of LDPC codes is accomplished with the Tanner graph-based algorithm. Decoding is achieved by passing messages along the lines of the graph. The messages on the lines that connect to the $i$th bit code $c_i$ are the estimates of $Pr\ (c_i = 1)$. The probabilities are combined in each of the check nodes in specific ways. The initial estimation probability of each bit node is the soft output of the channel. The bit node broadcasts this initial estimate to the parity check nodes on the lines connected to that bit node. Each check node makes new estimates for the bits involved in that parity equation and sends these new estimates back to the bit nodes. The probabilities of the message to be passed from bit node $c_i$ to the check node $f_j$ are described as follows:

$$q_{ij}(0) = 1 - p_i = \Pr(c_i = 0 \mid y_i) = \frac{1}{1 + e^{-2y_i/\sigma^2}}$$

$$q_{ij}(1) = p_i = \Pr(c_i = 1 \mid y_i) = \frac{1}{1 + e^{2y_i/\sigma^2}}$$

(10.8)

where $y_i$ is received signal. The probabilities of the message to be passed from check node $f_j$ to the bit node $c_i$ are given in Equation (10.9).

$$r_{ji}(0) = \frac{1}{2} + \frac{1}{2} \prod_{i' \in R_{j/i}} (1 - 2q_{i'j}(1))$$

$$r_{ji}(1) = 1 - r_{ji}(0)$$

(10.9)

where $R_{j/i}$ the set of column locations of 1 is in the $j$th row, excluding location $i$. At the next step, the probabilities of the message to be passed from bit node to the check node are

$$q_{ij}(0) = K_{ij}(1 - p_i) \prod_{j' \in C_{i/j}} r_{j'i}(0)$$

$$q_{ij}(1) = K_{ij} p_i \prod_{j' \in C_{i/j}} r_{j'i}(1)$$

(10.10)

where $C_{j/i}$ is the set of row locations of 1's in the $i$th column, excluding location $j$. $K_{ij}$ constants are chosen to ensure $q_{ij}(0) + q_{ij}(1) = 1$. The probabilities, which are used to determine if the received signal is a 0 or 1, are given in Equation (10.11).

$$Q_i(0) = K_i(1 - p_i) \quad \prod_{j \in C_i} r_{ij}(0)$$

$$Q_i(1) = K_i p_i \quad \prod_{j \in C_i} r_{ij}(1)$$

(10.11)

Here the constants $K_i$ are selected to ensure $Q_i(0) + Q_i(1) = 1$. After computing these probabilities, a hard decision is made as in Equation (10.12).

$$\hat{c}_i = \begin{cases} 1 & Q_i(1) \quad \dfrac{1}{2} \\ 0 & \text{other} \end{cases}$$

(10.12)

If $\hat{c}H^T = 0$ or the number of iterations exceeds limitations then computing the estimations is completed. If the necessary conditions are not supported, the probabilities in Equations (10.9)–(10.11) are computed again.

## 10.3 LDPC-Like Codes: GC Codes

Geometric construction (GC) codes are a new heuristic construction technique that generates all the even codes of length greater than eight with full rank property for Hamming distance of 4 [12]. The codes generated by the method in [12] include the extended Hamming codes and the Reed–Muller (RM) codes of distance of 4. The method for obtaining codes with a higher minimum distance using the proposed method is also described.

The goal of channel coding is finding a code that is easy to encode and decode, but at the same time gives a high code rate for the largest minimum distance [13]. The encoder and decoder implementation of long-length linear block codes is difficult in practice and also requires a large memory capacity. LDPC codes [14] receive enormous interest from the coding community as they hold higher BER performances [15]. They have a disadvantage where the encoding part is concerned because they have large, high-density, irregular generator matrices to encode information data, which cause difficulty in encoder implementations in practice. A generator matrix with a low density of ones is considered a low-density generator matrix (LDGM). It has the practical advantage of decreased encoding complexity and increased decoder architecture flexibility [16], and also performs close to the Shannon theoretical limit [17]. Block turbo codes (BTC) or turbo product codes (TPC) [18], and product accumulate (PA) codes [19] have been shown to provide near-capacity performance and low encoding and decoding complexity. They use simpler components codes to produce very long block codes. Recently [20], a similar approach to PA codes has been implemented using extended Hamming codes that achieve near Shannon limit performance for very high-rate coding, which is useful for bandwidth efficient transmission. On the other hand, it has some difficulty adjusting code rates when using extended Hamming codes as there are only limited numbers of code rates available with extended Hamming codes.

GC generates all the optimal Hamming distance of 4 even codes greater than eight. Here, by optimal Hamming distance of 4 code, a code of $C = (n, k, d)$ is meant, where $n$, $k$, and $d$ are length, dimension and minimum Hamming distance, respectively, has the full information rate for Hamming distance of 4 with respect to the table of best known codes in [9]. GC codes are denoted because of their geometric placement of component generator matrices in the construction. Extended Hamming codes and distance of 4 RM codes are subsets of GC construction codes because they generate codes as the power of 2 in length; whereas our GC codes generate the codes as the multiple of 2 in length. Therefore, our GC codes can be a good alternative for the component codes used in [7] and [8] because they give great flexibility and thus, adaptability with respect to the length of a code. Moreover, the constructed generator matrices of GC codes contain low density of ones and also have quasi-cyclic property in their component generator matrices. A code with cyclic or quasi-cyclic property can be encoded with less complexity [10,11] in a similar way of cyclic codes. Therefore, GC codes provide the advantages of LDGM codes and quasi-cyclic codes.

### 10.3.1 Hamming Distance 4 GC Codes

The ultimate binary generator matrices G of our GC codes are constructed using Equation (10.4) by geometrically placing the component generator matrices $G_1 = [1\ 1]$ and $G_2 = [1\ 0]$ of the codes $C_1 = (n_1, k_1, d_1) = (2,1,2)$ and $C_2 = (n_2, k_2, d_2) = (2,1,1)$, respectively. We formulate the size of GC codes as in (10.13).

$$C = (n, k, d) = \left(n, n - \lceil \log_2 n + 1 \rceil, 4\right) \tag{10.13}$$

where the code length, $n$, is an even number greater than or equal to 8, and $\lceil b \rceil$ notation rounds $b$ to the nearest integer value toward infinite.

$$
G = \begin{pmatrix}
\begin{matrix} G_1 G_1 \\ \quad G_1 G_1 \end{matrix} & & & & & & & & \\
& \cdot & \quad \cdot & \quad \cdot & & & & \begin{matrix} G_1 G_1 \\ \quad G_1 G_1 \end{matrix} & \\[2ex]
\begin{matrix} G_2 G_2 G_2 G_2 \\ \quad G_2 G_2 G_2 G_2 \end{matrix} & & & & & & & & \\
& & \cdot & \quad \cdot & \quad \cdot & & G_2 G_2 G_2 G_2 & & \\[2ex]
G_2 & G_2 & G_2 & G_2 & & & & & \\
& & G_2 & G_2 & G_2 & G_2 & & & \\
& & & \cdot & \cdot & \cdot & & & \\
& & & & G_2 & G_2 & G_2 & G_2 & \\
G_2 & & G_2 & & G_2 & & G_2 & & \\
& & G_2 & & G_2 & G_2 & & G_2 & \\
& & & & \cdot & \cdot & \cdot & & \\
& & \cdot & \quad \cdot & \quad \cdot & & & &
\end{pmatrix}
\left. \begin{matrix} \\ \\ \\ \end{matrix} \right\} E_1
\left. \begin{matrix} \\ \\ \\ \end{matrix} \right\} E_2
\left. \begin{matrix} \\ \end{matrix} \right\} E_3
\left. \begin{matrix} \\ \\ \end{matrix} \right\} E_4
\qquad (10.14)
$$

Then, G can be written as follows:

$$
G = \begin{bmatrix} D_1 \\ D_2 \\ \vdots \\ D_t \end{bmatrix}
\begin{matrix} \}E_1 \\ \}E_2 \\ \vdots \\ \}E_t \end{matrix}
\qquad (10.15)
$$

Here, $E_i$ ($i = 1, 2,\ldots,t$) represents group labels of component generator matrices with respect to their geometrical placement and they are represented by $D_i$ matrices, where $t$ is the number of the groups.

A set of simple geometrical placement rules to construct the generator matrices of GC codes using Equation (10.14) is as follows: Group the rows of G regarding their placement as $\{E_1, E_2,\ldots\}$. For $E_1$, place 2 of $G_1$s with one after another in the first row of G. The columns of the first row are shifted to the right by a scale of one $G_1$ and placed in the second row. This cyclic process is repeated for the following rows of group $E_1$ until the $G_1$ matrices arrive to the last column of a row. For $E_2$, place 4 of $G_2$s one after another in the first row of the group $E_2$. The columns of the first row are shifted to the right by a scale of 2 $G_2$s and placed in the second row of $E_2$. The process of shifting to the right and placing the $G_2$s matrices, which is also a cyclic process, is performed as long as there is room to place four $G_2$s matrices in the row of G. If there are not enough columns to place four $G_2$s, then stop placing $G_2$s. When the placement of the group $E_2$ is completed, start placing $G_2$s of group $E_3$, if there is room. In this case there will be one interval among four $G_2$s in a row. At the second row of $E_3$, the shifting to the right will be by four times the scale $G_2$. Similarly, this process is performed until there is no room left for placing four $G_2$s in a row. After the placement of $E_3$ is completed, if there is still room for placing $G_2$s of group $E_4$, four $G_2$s are placed with three intervals between them. In this group, shifting to the right interval will be by eight times the scale of $G_2$. This process continues for other groups {$E_4$, $E_5$, ...} and the whole process ends when there is no room left to place four $G_2$s in a row. In general, the shifting is 1 for the group $E_1$, 2 for the group $E_2$, four for the group $E_3$ and then becomes eight, sixteen and so on as the power of 2 for the following groups. The intervals among four $G_2$s start with 0 for the group $E_2$, and become 1 for the group $E_3$, and then 3, 7, 15 and so on for the following groups, which is calculated as half of the number of the shifting minus one.

Theorem 3 is given to prove that GC code construction generates the codes with minimum Hamming distance of 4. Before that some definitions are given and also Theorems 1 and 2 to constitute a base for Theorem 3.

If **u** and **v** are binary vectors, we let **u** **v** denote the number of 1's that **u** and **v** have in common. The following well-known equation

$$wt(\mathbf{u} + \mathbf{v}) = wt(\mathbf{u}) + wt(\mathbf{v}) - 2(\mathbf{u} \ \mathbf{v})$$

is very useful, where $wt(\mathbf{c})$ denotes Hamming weight of $c$.

**Theorem 1.** If the rows of a generator matrix G for a binary $(n, k)$ code C have even weight and are orthogonal to each other, then C is self-orthogonal [12].

**Theorem 2.** If the rows of a generator matrix G for a binary $(n, k)$ code C have weights divisible by 4 and are orthogonal to each other, then C is self-orthogonal and all weights in $C$ are divisible by 4 [12].

**Lemma 1.** A codeword of a code, $C_i$, generated by a group generator matrix, $D_i$, of Equation (10.15) has the Hamming weight of at least 4.

**Proof.** Rows of a group generator matrix, $D_i$ of Equation (10.15) have weight 4, namely even, and the rows of a $D_i$ are orthogonal to each other. In each group the orthogonality property holds. This provides the conditions of Theorem 1, therefore the code $C_i$ generated by $D_i$ is self-orthogonal. Because the weights of the rows of $D_i$ in Equation (10.15) are 4, namely divisible by 4 and the $C_i$ is self-orthogonal, then, by Theorem 2, all weights in the code $C_i$ generated by $D_i$ are divisible by 4. Each row of $D_i$ has a weight 4 divisible by 4, hence the minimum Hamming weight or the minimum Hamming distance of a code $C_i$ is 4. Lemma 1 proves that component codes of each group in Equations (10.15) and (10.14) generate the code words with a minimum Hamming distance of 4.

**Lemma 2.** If $\mathbf{u}$ and $\mathbf{v}$ are distinct binary row vectors of the G in Equation (10.14), then $\mathbf{u} \cdot \mathbf{v}$ is at most 2.

**Proof.** The geometric placement rules of the G naturally provide this, as the number of overlapping component matrices among the rows of G is at most 2, which can easily be observed from the G in Equation (10.14).

***Theorem 3.*** A code, C, generated by the matrix, G, of Equation (10.14) has the minimum Hamming distance of 4.

**Proof.** Let $\mathbf{u}$ and $\mathbf{v}$ be the distinct binary row vectors of G in Equation (10.14), and we know $wt(\mathbf{u} + \mathbf{v}) = wt(\mathbf{u}) + wt(\mathbf{v}) - 2(\mathbf{u} \cdot \mathbf{v})$. By Lemma 1, $wt(\mathbf{u}) \geq 4$ and $wt(\mathbf{v}) \geq 4$. By Lemma 2, $\mathbf{u} \cdot \mathbf{v} \leq 2$. Then, the worst-case scenario is $wt(\mathbf{u} + \mathbf{v}) = 4 + 4 - 2*2 = 4$. Hence the minimum Hamming distance is 4.

Theorem 3 is verified by conducting an exhaustive computer search and also formulated the size of verified codes as in Equation (10.13) to generalize the constructed optimal codes for Hamming distance of 4. The Equation (10.13) is an important property of Equation (10.14) that allows controlling the size of constructed codes. For instance, if we want to construct a distance of 4 binary linear block code of length $n = 1200178$, then, $k$ is equal to 1200156, which means we can have a code $C = (1200178, 1200156, 4)$ using Equation (10.14). This flexibility property of adjusting a code length arbitrarily is important for many applications.

For example in [8], a code of length 256 is constructed using one (128,120) and two (64,57) extended Hamming codes and the resulting code rate becomes R = (120 + 57*2)/(128 + 64*2) = 0.914. The same length code using GC construction is in the size of (256, 247, 4) and the code rate R = 247/256 = 0.964. Therefore, this code construction provides 0.05 better code rate. This example shows only a better coding rate of this GC construction; in addition, it is also advantageous with respect to the flexibility in adjusting the length of a code.

## 10.3.2 Higher-Distance GC Codes

Hamming distance of 8 GC codes can also be constructed using Equation (10.14) by choosing $G_1 = [1\ 1\ 1\ 1]$ and $G_2 = [1\ 1\ 0\ 0]$; however, the codes rate may not be optimal. To improve code rate we use another component generator matrix $G_3 = [0\ 1\ 1\ 0]$ and place it exactly the same way we do with $G_2$ after finishing the placement of $G_2$. For Hamming distance of 16 codes, we choose $G_1 = [1\ 1\ 1\ 1\ 1\ 1\ 1\ 1]$, $G_2 = [1\ 1\ 1\ 1\ 0\ 0\ 0\ 0]$, $G_3 = [0\ 0\ 1\ 1\ 1\ 1\ 0\ 0]$, and $G_4 = [1\ 0\ 1\ 0\ 1\ 0\ 1\ 0]$, where $G_2$, $G_3$, and $G_4$ are placed in the same manner.

### 10.3.3 Simulation Results

Some simulation results are shown in Figure (10.2) to assess the BER performances of GC codes. The error performance of the (14,8) extended LDPC code [13] over the AWGN channel using a sum-product algorithm (SPA) [14] with maximum of 200 iteration, as well as the uncoded BPSK over the AWGN channel for comparison purposes have been plotted. The considered GC codes have been decoded using SPA with a maximum of 40 iterations. In all the considered codes, binary phase shift keying (BPSK) modulation is used and the perfect bit and block synchronizations are assumed.

As an example, GC (12,7,4) BER performance is similar to (14,8) extended LDPC code of [13] in AWGN channel. Also, GC (12,7,4) provides about 2.2 dB coding gain in AWGN channel compared to the uncoded case. These simulations demonstrate that GC codes have similar performance to their counterparts, but they have advantages from the implementation point of view because they have the properties of the generator matrices such as quasi-cyclic, regular, flexible in length, controllable size, and low-density.

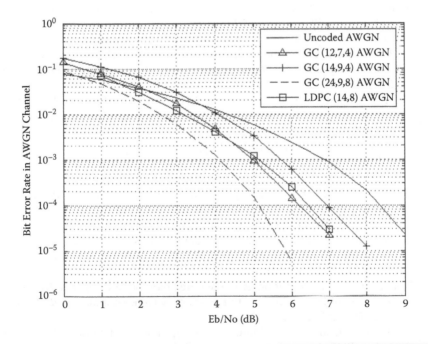

**Figure 10.2   BER of some GC codes, an extended LDPC code and uncoded BPSK modulation. (Copyright permission by Elsevier [12].)**

# References

[1] Gallager, R. G. *Low-Density Parity-Check Codes*. Cambridge, MA: MIT Press, 1963.

[2] Zyablov, V., and M. Pinkster. Estimation of the Error-Correction Complexity of Gallager Low-Density Codes. *Probl. Pered. Inform.* 11:23–26, Jan. 1975.

[3] Margulis, G. A. Explicit Construction of Graphs without Short Cycles and Low Density Codes. *Combinatorica* 2(1):71–78, 1982.

[4] Tanner, R. A Recursive Approach to Low Complexity Codes. *IEEE Trans. Inform. Theory* IT-27:533–547, Sept. 1981.

[5] MacKay, D. J. C., and R. M. Neal. Near Shannon Limit Performance of Low Density Parity Check Codes. *Electron. Lett.* 32:1645–1646, Aug. 1996.

[6] Wiberg, N. Codes and Decoding on General Graphs. Ph.D. Thesis 440, Dept. Elect. Linköping Univ., Linköping, Sweden, 1996.

[7] Luby, M. G., M. Mitzenmacher, M. A. Shrokrollahi, D. Spielman, and V. Stemann. Practical Loss-Resilient Codes. *Proc. 29th Annu. ACM Symp. Theory of Computing* 150–159, 1997.

[8] Luby, M. G., M. Mitzenmacher, M. A. Shrokrollahi, D. Spielman, and V. Stemann,. Practical Loss-Resilient Codes. *IEEE Trans. Inform. Theory* 47: 569–584, Feb. 2001.

[9] Rimoldi, B. E. A Decomposition Approach to CPM. *IEEE Trans. on Inform. Theory* 34:260–270, Mar. 1988.

[10] Rimoldi, B. E. Design of Coded CPFSK Modulation Systems for Band-Width and Energy Efficiency. *IEEE Trans.on Commun.* 37:897–905, Sept. 1989.

[11] Hekim, Y, N. Odabasioglu, and O. N. U•an. Performance of Low Density Parity Check/Continuous Phase Frequency Shift Keying (LDPC/CPFSK) over Fading Channels. *International Journal of Communication Systems* 20(4):397–410, 2007.

[12] Altay, G., and O. N. U•an. Heuristic Construction of High-Rate Linear Block Codes. *International Journal of Electronics and Communication, AEU* 663–666, 2006.

[13] Bossert, M. *Channel Coding*. Wiley, 1999.

[14] Gallager, R. *Low-Density Parity-Check Codes*. Cambridge, MA: MIT Press, 1963.

[15] Chung, S.Y., G. D. Forney, T. J. Richardson, and R. Urbanke. On the Design of Low-Density Parity-Check Codes within 0.0045 dB of the Shannon Limit. *IEEE Comm. Letter* 5(2):58–60, Feb. 2001.

[16] Oenning, T. R., and J. Moon. A Low-Density Generator Matrix Interpretation of Parallel Concatenated Single Bit Parity Codes. *IEEE Trans. Magn.* 37(2):737–741, Mar. 2001.

[17] Frias, J. G., and W. Zhong. Approaching Shannon Performance by Iterative Decoding of Linear Codes with Low-density Generator Matrix. *IEEE Comm. Letter* 7(6):266–268, June 2003.

[18] Pyndiah, R. M. Near-optimum Decoding of Product Codes: Block Turbo Codes. *IEEE Trans. Commun.* 46(8):1003–1010, Aug. 1998.

[19] Li, J., K. R. Narayanan, and C. N. Georghiades. Product Accumulate Codes: A Cass of Codes with Near-capacity Performance and Low Decoding Complexity. *IEEE Trans. Inform. Theory* 50:31–46, Jan. 2004.

[20] Isaka, M., and M. Fossorier. High-Rate Serially Concatenated Coding with Extended Hamming Codes. *IEEE Comm. Letter* 9(2):160–162, Feb. 2005.

[21] Brouwer, A. E., and T. Verhoeff. An Updated Table of Minimum-distance Bounds for Binary Linear Codes. *IEEE Trans. Inform. Theory* 39(2):664–677, March 1993.

[22] Lucas, R., M. P. C. Fossorier, Y. Kou, and S. Lin. Iterative Decoding of One-Step Majority Logic Decodable Codes based on Belief Propagation. *IEEE Trans. Commun.* 48:931–937, June 2000.

[23] Jhonson, S. J., and S. R. Weller. A Family of Irregular LDPC Codes with Low Encoding Complexity. *IEEE Comm. Letter* 7(2)79–81, Feb. 2003.

[24] Pless, V. *Introduction to the Theory of Error-Correcting Codes, 2nd edition.* John Wiley and Sons, Inc., 1989.

[25] Lin, S., and D. J. Costello, Jr. *Error Control Coding, Second edition.* Upper Saddle River, NJ: Prentice Hall, 2004.

[26] MacKay, D. Good Error Correcting Codes Based on Very Sparse Matrices. *IEEE Trans. Inform. Theory* 399–431, Mar. 1999.

# Chapter 11

# Coding Techniques Using Continuous Phase Modulation

In band-limited channels, such as deep space and satellite communications, continuous phase modulation (CPM) has explicit advantages, because it has a low spectral occupancy property. CPM is composed of the continuous phase encoder (CPE) and memoryless mapper (MM). CPE is a convolutional encoder producing code word sequences that are mapped onto waveforms by the MM, creating a continuous phase signal. CPE-related schemes have better performance than systems using the traditional approach for a given number of trellis states, because they increase Euclidean distance. Once the memory structure of CPM is assigned, it is possible to design a single-joint convolutional code, composed of trellis- and convolutionally-coded CPM systems as in [1–3].

Rimoldi derived the tilted-phase representation of CPM in [1], with the information bearing phase given by

$$\phi(t, Y) = 4\pi h \sum_{i=0}^{\infty} Y_i \, q(t - iT) \tag{11.1}$$

The modulation index $h$ is equal to $J/P$, where $J$ and $P$ are relatively prime integers. $Y$ is an input sequence of M-ary symbols, $Y_i \in \{0, 1, 2, \ldots, \text{M-1}\}$. $T$

**243**

is the channel symbol period. $J$ is generally chosen as 1 and the modulation index appears in the form of $h = 1/P$. $P$ is a number that can be calculated as 2 to the power of $\lambda$, number of memories in CPE. The phase response function $q(t)$, is a continuous and monotonically increasing function subject to the constraints

$$q(t) = \begin{cases} 0 & t \le 0 \\ 1/2 & t \ge LT \end{cases} \tag{11.2}$$

where $L$ is an integer. The phase response is usually defined in terms of the integral of a frequency pulse $g(t)$ of duration $LT$, i.e.,

$$q(t) = \int_{-\infty}^{t} g(\tau) d\tau$$

For full response signaling $L$ equals 1, and for partial response systems, $L$ is greater than 1. Finally, the transmitted signal $s(t)$ is as,

$$s(t,Y) = \sqrt{\frac{2E_s}{T}} \cos(2\pi f_1 t + \phi(t,Y) + \phi_0) \tag{11.3}$$

where $f_1$ is the asymmetric carrier frequency, as $f_1 = f_c - h(M-1)/2T$, and $f_c$ is the carrier frequency. $E_s$ is the energy per channel symbol and $\phi_0$ is the initial carrier phase. We assume that $f_1 T$ is an integer; this condition leads to a simplification when using the equivalent representation of the CPM waveform.

## 11.1 Multilevel Turbo–Coded Continuous Phase Frequency Shift Keying (MLTC-CPFSK)

The challenge of finding practical decoders for large codes was not addressed until turbo codes were introduced by Berrou et al. in 1993 [4]. The performance of these new codes is close to the Shannon limit with relatively simple component codes and large interleaver. Turbo codes are a new class of error correction codes that were introduced along with a practical decoding algorithm. The importance of turbo codes is that they enable reliable communications with power efficiencies close to the theoretical limit predicted by Claude Shannon. Turbo codes are the most efficient codes for low-power applications such as deep space and satellite communications, as well as for interference-limited applications such as third-generation cellular and personal communication services.

In trellis-based structures, to improve the bit-error probability, many scientists not only study the channel parameters as in [5,6] but as in [7–10] they have also used multilevel coding as an important band and power efficient technique because it provides a significant amount of coding gain and low coding complexity. A multilevel encoder is a combination of several error correction codes applied to subsets of some signal constellation. The multilevel coding scheme employs, at each signaling interval, one or more output bits of each of several binary error-control encoders to construct the signal to be transmitted. An important parameter of a coded modulation scheme is the computational complexity of the decoder. Usually, a kind of suboptimal decoder, called the multistage decoder, is used for multilevel codes [8–14]. Currently there are also many attempts to improve the performance of multilevel turbo-based systems. In [15] the author discusses the impact of interleaver size on the performance of multilevel turbo-based systems. Power and bandwidth efficiencies of multilevel turbo codes are also discussed in [16,17].

To improve error performance and bandwidth efficiency, multilevel turbo coding and continuous-phase modulation are combined in [15,16] and multilevel turbo coded continuous-phase frequency shift keying (MLTC-CPFSK) is introduced. In a MLTC-CPFSK transmitter, each turbo encoder is defined as a level. Because the number of encoders and decoders are the same for each level of the multilevel encoder, there is a corresponding decoder, defined as a stage. Furthermore, except for the first decoding stage, the information bits, which are estimated from the previous decoding stages, are used for the next stages. This process is known as multistage decoding (MSD). The basic idea of multilevel coding is to partition a signal set into several levels and to encode separately each level by a proper component of the encoder. In multilevel turbo coded continuous-phase modulation (MLTC-CPM), to provide phase continuity of the signals, turbo encoder and continuous phase encoder (CPE) are serially concatenated at the last level [17–19], while all other levels consist of only a turbo encoder (Figure 11.1). Therefore, the system contains more than one turbo encoder/decoder blocks in its structure. The parallel input data sequences are encoded and mapped to continuous-phase frequency shift keying (CPFSK) signals. Then these modulated signals are passed through AWGN and fading channels. At the receiver side, the BW algorithm is used to estimate the channel parameters before the input sequence of first level is estimated from the first turbo decoder and the other input sequences are computed by using the estimated input bit streams of previous levels.

## 11.1.1 Encoder Structure

The multilevel turbo CPFSK encoder and decoder consist of many parallel turbo encoder/decoder levels as in Figure 11.1. There is one binary turbo

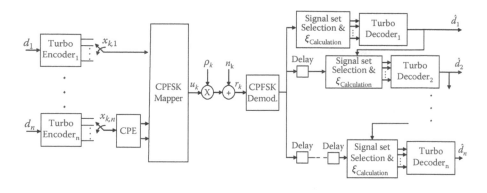

**Figure 11.1  MLTC-CPFSK block diagram. (Copyright permission by John Wiley & Sons [15].)**

encoder at every level of the multilevel turbo encoder and a CPE is placed after the last level turbo encoder. Here, CPE is used to achieve the phase continuity of the transmitted signals and its state indicates the initial phase of the instant signal that is given in (11.3) as $\varphi_0$. Each turbo encoder is fed from one of the input bit streams, which are processed simultaneously. The outputs of these encoders can be punctured and thereafter only the last level output is passed through CPE. Then, these outputs are mapped to CPFSK signals according to the partitioning rule. In partitioning, $x_{k,1}$ is the output bit of the first-level turbo encoder where the signal set is divided into two subsets. If $x_{k,1} = 0$, then the first subset is chosen, if $x_{k,1} = 1$, then the second subset is chosen. The $x_{k,2}$ bit is the output bit of the second-level turbo encoder and divides the subsets into two just as in previous levels. This partitioning process continues until the last partitioning level has occurred. At the last level, to provide phase continuity, the CPE encoder is placed after last-level turbo encoder. Therefore, at this level, the signal set is divided twice and hence the signal, which will be sent to channel, is selected. In Figure 11.2, as an example, the multilevel turbo coding system for 4-ary CPFSK-2 level turbo codes (2LTC-CPFSK) is given.

In each level, we consider a one-third recursive systematic convolutional (RSC) encoder with memory size two. For each level, input bit streams are encoded by the turbo encoders. At turbo encoder outputs, the encoded bit streams to be mapped to M-ary CPFSK signals are determined after a puncturing procedure. The first bit is taken from the first-level turbo encoder output, the second bit is taken from second-level encoder output and the other bits are obtained similarly. Following this

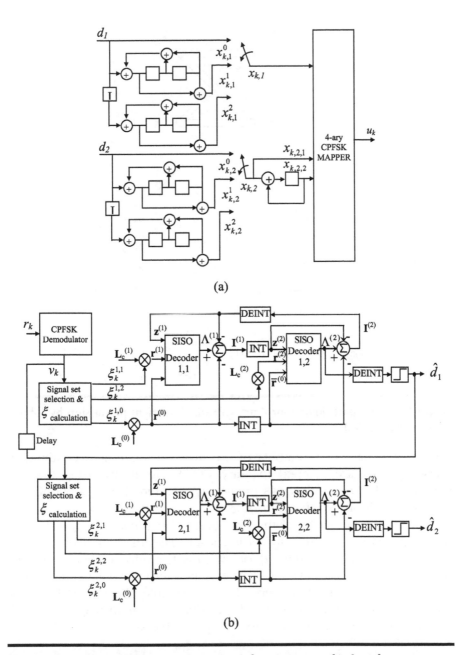

**Figure 11.2  2LTC-CPFSK system. (a) Encoder structure; (b) decoder structure. (Copyright permission by John Wiley & Sons [15].)**

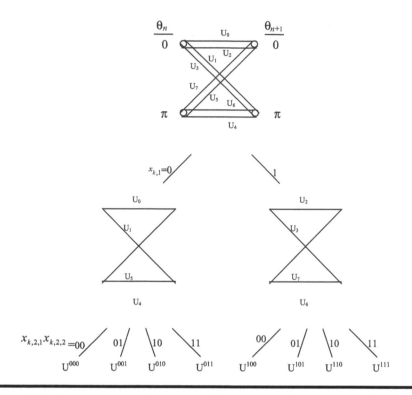

**Figure 11.3** **Set partitioning for 4-ary CPFSK. (Copyright permission by John Wiley & Sons [15].)**

process, the bits at the outputs of the turbo encoders and continuous-phase encoder are mapped to 4-ary CPFSK signals by using the set-partitioning technique that was previously mentioned. Set partitioning of 2LTC-CPFSK is shown in Figure 11.3.

Here, if the output bit of the first-level turbo encoder is $x_{k,1} = 0$, then $u_0^1$ set, if it is $x_{k,1} = 1$, then $u_1^1$ set is chosen. The first output bit $\{x_{k,2,1}\}$ of the CPE determines whether $u_1^2$ or $u_0^2$ subsets are to be chosen and the second output bit $\{x_{k,2,2}\}$ of the CPE selects the signal that will be transmitted using previous partitioning steps. In this example, the remainder $w_k$ of the encoder can be found using the feedback polynomial $g^{(0)}$ and feedforward polynomial is $g^{(1)}$. The feedback variable is

$$w_k = d_k + \sum_{j=1}^{K} w_{k-j} g_j^{(0)} \tag{11.4}$$

and the RSC encoder outputs $x_k^{1,2}$, that are called parity data is

$$x_{k,}^{\lambda} = \sum_{j=0}^{K} w_{k-j} g_j^{(1)} \qquad (11.5)$$

where $\lambda$ indicates the number of half recursive systematic encoders for each level of MLTC encoder and shows the level number of the overall MLTC encoder.

In Figure 11.2, the RSC encoder has the feedback polynomial $g^{(0)} = 7$ and a feedforward polynomial $g^{(1)} = 5$, and it has a generator matrix. Here $D$ is the memory unit.

$$G(D) = \begin{bmatrix} 1 & \dfrac{1 + D + D^2}{1 + D^2} \end{bmatrix} \qquad (11.6)$$

As another example, 3LTC-CPFSK is shown as in Figure 11.4, to clarify the general structure of MLTC-CPFSK scheme.

## 11.1.2 Decoder Structure

The problem of decoding turbo codes involves the joint estimation of two Markov processes, one for each constituent code. Although in theory it is possible to model a turbo code as a single Markov process, such a representation is extremely complex and does not lend itself to computationally tractable decoding algorithms. In turbo decoding, there are two Markov processes, defined by the same set of data; hence, the estimation can be refined by sharing the information between the two decoders in an iterative fashion. More specifically, the output of one decoder can be used as *a priori* information by the other decoder. If the outputs of the individual decoders are in the form of hard-bit decisions, then there is little advantage in sharing information. However, if the individual decoders produce soft-bit decisions, considerable performance gains can be achieved by executing multiple iterations of decoding. The received signal can be shown as

$$r_k = \rho_k u_k + n_k \qquad (11.7)$$

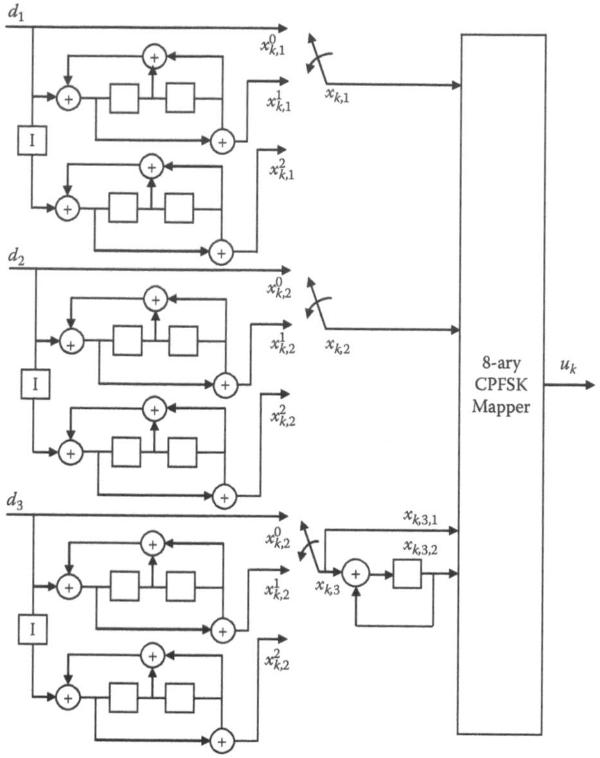

(a)

**Figure 11.4    3LTC-CPFSK system. (a) Encoder structure; (b) decoder structure. (Copyright permission by John Wiley & Sons [15].)**

where $r_k$ is a noisy received signal, $u_k$ is the transmitted MLTC-CPFSK signal, $\rho_k$ is the fading parameter and $n_k$ is Gaussian noise at time $k$. The maximum *a posteriori* (MAP) algorithm calculates the *a posteriori* probability of each bit.

Let $\tilde{\gamma}^{st}(s_k \rightarrow s_{k+1})$ denote the natural logarithm of the branch metric $\gamma^{st}(s_k \rightarrow s_{k+1})$, where $s_k$ is the encoder state at $k$th coding step, $st$ is the decoding stage, then

$$\tilde{\gamma}^{st}(s_k \rightarrow s_{k+1}) = \ln \gamma^{st}(s_k \rightarrow s_{k+1}) \tag{11.8}$$

$$\ldots = \ln P\left[d_k\right] + \ln P\left[r_k \left| x_k \right.\right]$$

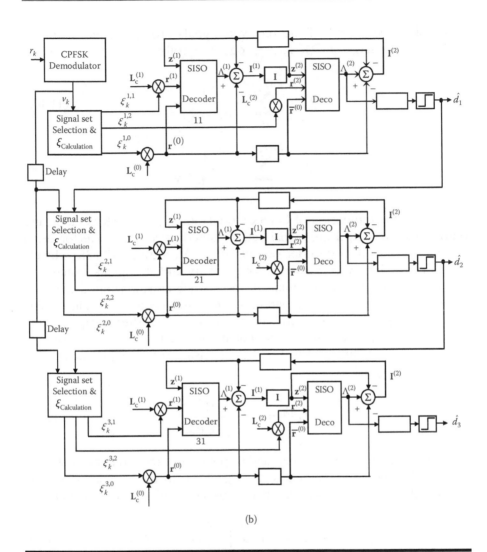

(b)

**Figure 11.4 (continued)** **3LTC-CPFSK system. (a) Encoder structure; (b) decoder structure. (Copyright permission by John Wiley & Sons [15].)**

where

$$\ln P\left[d_k\right] = z_k d_k - \ln(1 + e^{z_k}) \tag{11.9}$$

$z_k$ is the *a priori* information that is obtained from the output of the other decoder. For every decoding stage of MLTC-CPFSK, zero and one probabilities $\{P_{k,0}^{st}, P_{k,1}^{st}\}$ of the received signals are calculated at time $k$ and decoding stage $st \in \{1, 2, \ldots, \log_2 M\}$ as follows:

$$P_{k,0}^{st} = \sum_{j=0}^{(M/st)-1} \frac{1}{(v_k - u_{0,j}^{st})^2} \qquad (11.10a)$$

$$P_{k,1}^{st} = \sum_{j=0}^{(M/st)-1} \frac{1}{(v_k - u_{1,j}^{st})^2} \qquad (11.10b)$$

where $M$ is the dimension of CPFSK modulation, $v_k$ is CPFSK demodulator output and $\{u_{0,j}^{st}, u_{1,j}^{st}\}$ are signal sets obtained by the set selector using previous stage turbo decoder output $\{\hat{d}_{st-1}\}$. In a MLTC-CPFSK scheme, each digit of binary correspondence of M-ary CPFSK signals matches to one stage from the most significant to least significant while stage level $st$ increases. The signal set is partitioned into subsets for each binary digit matching stage, depending on whether it is 0 or 1. After computing the one and zero probabilities as in Equations (11.10a) and (11.10b), the received signal is mapped to $\{-1,1\}$ range using 0 and 1 probabilities of the received signal as

$$\xi_k^{st,q} = 1 - \frac{2 \cdot P_{k,0}^{st}}{P_{k,0}^{st} + P_{k,1}^{st}} \qquad (11.11)$$

These probability computations and mapping are executed in every stage of the decoding process according to the signal set. In the decoder scheme in Figure 11.1, the signal set selector operates using Equations (11.11) and (11.12).

$$\tilde{\gamma}^{st}(s_k \rightarrow s_{k+1}) = \ln P[d_k] - \frac{1}{2}\ln(\pi N_0 / E_s) - \qquad (11.12)$$

$$\frac{E_s}{N_0} \sum_{q=0}^{n-1} \left[ \xi_k^{st,q} - (2x^q - 1) \right]^2$$

Now let $\tilde{\alpha}^{st}(s_k)$ be the natural logarithm of $\alpha^{st}(s_k)$,

$$\tilde{\alpha}^{st}(s_k) = \ln \alpha^{st}(s_k)$$

$$= \ln \left\{ \sum_{s_{k-1} \in A} \exp\left[ \tilde{\alpha}^{st}(s_{k-1}) + \tilde{\gamma}^{st}(s_{k-1} \rightarrow s_k) \right] \right\} \qquad (11.13)$$

where $A$ is the set of states $s_{k-1}$ that are connected to the state $s_k$. Now let $\tilde{\beta}^{st}(s_k)$ be the natural logarithm of $\beta^{st}(s_k)$,

$$\tilde{\beta}^{st}(s_k) = \ln \beta^{st}(s_k) \tag{11.14}$$

$$\dots = \ln\left\{ \sum_{s_{k+1} \in B} \exp\left[ \tilde{\beta}^{st}(s_{k+1}) + \tilde{\gamma}^{st}(s_k \to s_{k+1}) \right] \right\}$$

where $\beta$ is the set of states $s_{k+1}$ that are connected to state $s_k$, and we can calculate the log likelihood ratio (LLR) by using

$$\Lambda_k^{st} = \ln \frac{\sum_{S_1} \exp\left[ \tilde{\alpha}^{st}(s_k) + \tilde{\gamma}^{st}(s_k \to s_{k+1}) + \tilde{\beta}^{st}(s_{k+1}) \right]}{\sum_{S_0} \exp\left[ \tilde{\alpha}^{st}(s_k) + \tilde{\gamma}^{st}(s_k \to s_{k+1}) + \tilde{\beta}^{st}(s_{k+1}) \right]} \tag{11.15}$$

where $S_1 = \{s_k \to s_{k+1} : d_k = 1\}$ is the set of all state transitions associated with a message bit of 1, and $S_0 = \{s_k \to s_{k+1} : d_k = 0\}$ is the set of all state transitions associated with a message bit of 0. At the last iteration we make the hard decision by using the second decoder output $\Lambda(st,2)$,

$$\hat{d}_k = \begin{cases} 1 & \text{if } \Lambda^{st} 2 \geq 0 \\ 0 & \text{if } \Lambda^{st} 2 < 0 \end{cases} \tag{11.16}$$

### 11.1.3 Simulation Results over Rician Fading Channel

The BER versus SNR curves of the two-level turbo coded 4-ary CPFSK system and three-level turbo coded 8-ary CPFSK system are obtained for AWGN, Rician (for Rician channel parameter $K = 10$ dB) and Rayleigh channels via computer simulations with ideal channel state information (CSI). The results are shown in Figures 11.5–11.8. The frame sizes are chosen as 100 and 1024. To compare the scheme's performance, two well-known best codes are referred to that are presented in [20] by Naraghi-Pour. These reference codes are called Ref-1 and Ref-2. Ref-1 is a binary trellis coded 4-ary CPFSK scheme with $R_s = 2/3$ coding rate. The first example, the 2LTC-CPFSK system, compares well with Ref-1, because they both have 4-ary CPFSK with $R_s = 2/3$. In the same way, Ref-2 is comparable with our second example system, 3LTC-CPFSK. Ref-2 and 3LTC-CPFSK systems use 8-ary

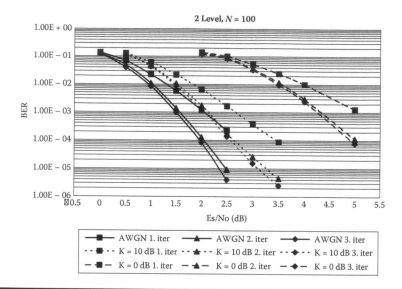

**Figure 11.5** Performances of 2LTC-CPFSK system for $N = 100$. (Copyright permission by John Wiley & Sons [15].)

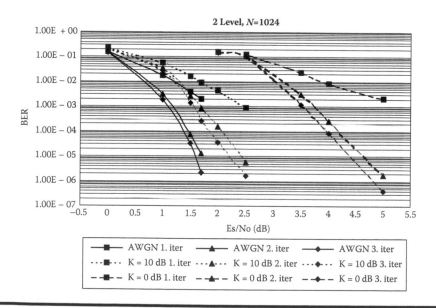

**Figure 11.6** Performances of 2LTC-CPFSK system for $N = 1024$. (Copyright permission by John Wiley & Sons [15].)

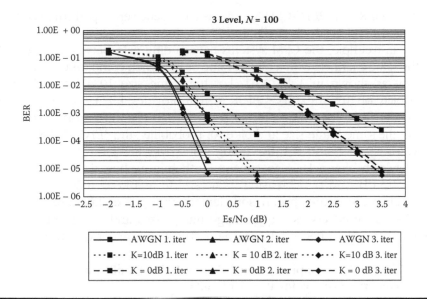

**Figure 11.7    Performances of 3LTC-CPFSK system for** $N$ **= 100. (Copyright permission by John Wiley & Sons [15].)**

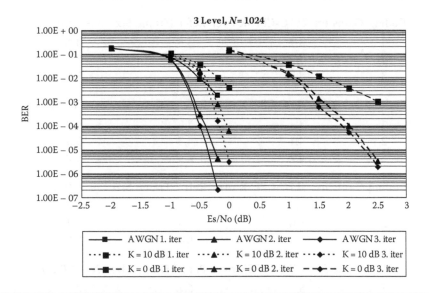

**Figure 11.8    Performances of 3LTC-CPFSK system for** $N$ **= 1024. (Copyright permission by John Wiley & Sons [15].)**

CPFSK with $R_s = 1$. 2LTC-CPFSK systems have better error performance than Ref-1 and Ref-2 in all channels and SNR values.

As an example, the CPM systems have coding gain between 3.1–8.3 dB for the same channels with a bit error rate of $10^{-4}$ when compared to reference systems as shown in Table 11.1. For the frame size 1024, at a bit error probability of $10^{-4}$, the 2LTC-CPFSK system provides 4.2, 4.4, 5.7 dB coding gains over Ref-1 for AWGN, Rician and Rayleigh channels, respectively.

For the same frame size and bit error probability, the 3LTC-CPFSK system provides 5.9, 6.9, and 8.3 dB coding gains over Ref-2 for AWGN, Rician, and Rayleigh channels, respectively. Furthermore, there is approximately only a 0.5 dB gain by increasing frame size from 100 to 1024. Thus, even if small frame sizes are chosen, sufficient bit error probabilities are obtained.

Classical turbo codes require large frame sizes and a high number of iterations to obtain better bit error rates. These two major disadvantages of

**Table 11.1  Example Systems' Coding Gains (in dB) over Reference Systems for $P_b = 10^{-4}$ 1024**

| Iteration | AWGN Channel | Rician Channel (K = 10 dB) | Rayleigh Channel |
|-----------|--------------|----------------------------|------------------|
| **Coding Gains for 2LTC-CPFSK over Ref-1 (Frame Size N = 100)** | | | |
| 1 | 3.1 | 3.5 | 4.1 |
| 2 | 3.7 | 3.9 | 5 |
| 3 | 3.8 | 4 | 5.1 |
| **Coding Gains for 2LTC-CPFSK over Ref-1 (Frame Size N = 1024)** | | | |
| 1 | 3.3 | 3.6 | 4.2 |
| 2 | 4.1 | 4.3 | 5.5 |
| 3 | 4.2 | 4.4 | 5.7 |
| **Coding Gains for 3LTC-CPFSK over Ref-2 (Frame Size N = 100)** | | | |
| 1 | 4.9 | 5.5 | 6.4 |
| 2 | 5.6 | 6.3 | 7.4 |
| 3 | 5.7 | 6.4 | 7.5 |
| **Coding Gains for 3LTC-CPFSK over Ref-2 (Frame Size N = 1024)** | | | |
| 1 | 5 | 6.1 | 7 |
| 2 | 5.8 | 6.8 | 8.1 |
| 3 | 5.9 | 6.9 | 8.3 |

*Source:* Osman, *Journal of IEE*, Vol. 20, No. 1, pp. 103–119, 2007.

turbo codes are reduced by MLTC-CPFSK schemes. As shown in Figures 11.5–11.8 and Table 11.1, the MLTC-CPFSK provides a significant amount of coding gain with short frame sizes and a few iterations. While the second iteration provides between 0.6 and 1.2 dB coding gains over the first iteration, the third iteration provides only 0.1–0.2 dB coding gain over the second iteration. Therefore, two iterations are adequate. Thus, the combined system decreases the iteration number and provides less sensitivity to frame sizes.

### *11.1.4  Simulation Results over MIMO Channel*

It is very important to know the channel capacity of the MLTC-CPFSK for multiple transmitter $(T_x)$ and multiple receiver $(R_x)$ antenna configurations as in Figure 11.9.

In CPFSK, the signal constellation consists of $V = 2M$ signals; there are eight different signals $\{s^0, s^1, s^2, s^3, -s^0, -s^1, -s^2, -s^3\}$ for 4CPFSK, where $M$ is the number of encoder outputs. In the AWGN case,

$$y(n) = s(n) + w(n) \tag{11.17}$$

where $s$ denotes a real valued discrete channel signal transmitted at modulation time $nT$, and $w$, is an independent, normally distributed noise sample with zero mean and variance $\sigma^2$ along each dimension, and $y$ is AWGN output. The average SNR is defined as

$$SNR = \frac{E\left\{ \left| s(n)^2 \right| \right\}}{E\left\{ \left| w(n)^2 \right| \right\}} = \frac{E\left\{ \left| s(n)^2 \right| \right\}}{M\sigma^2} \tag{11.18}$$

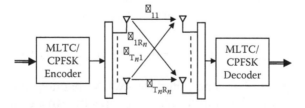

**Figure 11.9  General structure of MLTC-CPFSK for MIMO channels. (Copyright permission by John Wiley & Sons [15].)**

Normalized average signal power $E\left\{ \left| s(n)^2 \right| \right\} = 1$ is assumed. For fading channels, the received signal is

$$y(n) = \rho(n)s(n) + w(n) \tag{11.19}$$

where $\rho$ is the fading amplitude and has Rician distribution which is given below:

$$P(\rho) = 2\rho(1 + K)\ e^{(-\rho^2(1 + K) - K)} I_0 \left[ 2\ \rho\sqrt{K(1 + K)} \right] \tag{11.20}$$

$K$ is SNR of the fading effect of the channel and $I_0$ is the modified Bessel function of the first kind, order zero. If $K$ is assumed to be 0 dB, the channel becomes Rayleigh.

As a special case of Figure 11.9, for the 2Tx-1Rx configuration, the transmitted signals from each antenna at the same time interval are the same, thus, the received signal is

$$y(n) = \rho_{11}(n)s(n) + \rho_{21}(n)s(n) + w(n)$$
$$= (\rho_{11}(n) + \rho_{21}(n))s(n) + w(n) \tag{11.21}$$

Here, $\rho_{11}$ and $\rho_{21}$ are fading amplitudes for each multipath channels. For the 2Tx-2Rx configuration, received signals at received antennas are as follows:

$$y_1(n) = (\rho_{11}(n) + \rho_{21}(n))s(n) + w_1(n) \tag{11.22}$$

$$y_2(n) = (\rho_{12}(n) + \rho_{22}(n))s(n) + w_2(n) \tag{11.23}$$

Here, $y_1$ and $y_2$ are the received signals of the first and second receiver antennas. $\rho_{11}$, $\rho_{21}$, $\rho_{12}$, and $\rho_{22}$ are fading amplitudes for each multipath channels. $w_1$ and $w_2$ are noise values. The total received signal is

$$y(n) = (\rho_{11}(n) + \rho_{21}(n) + \rho_{12}(n) + \rho_{22}(n))s(n) + w_1(n) + w_2(n) \tag{11.24}$$

Channel capacities are computed for AWGN channels; therefore, fading parameters are considered as 1. Channel capacity can be calculated from the formula given in [18], if the codes with equal probability occurrence of channel input signals are of interest.

$$C^* \approx \log_2(V) - \frac{1}{V} \cdot \sum_{k=0}^{V-1} E \left\{ \log_2 \sum_{i=0}^{V-1} \exp \left[ -\frac{\left| s^k + w - s^i \right|^2 - \left| w \right|^2}{2\sigma^2} \right] \right\} \quad (11.25)$$

in bit/T. Using a Gaussian random number generator, $C^*$ has been evaluated by Monte Carlo averaging of Equation (11.25). In Figure 11.9, channel capacity is plotted as a function of SNR for different dimensional CPFSK.

In Figure 11.10, the channel capacities of 2LTC-4CPFSK, $h = 1/2$, are shown for different antenna configurations. The maximum channel capacity is 3, because $V = 2M = 8$ and $C^*_{max} = \log_2(8)$. In that case, for 4-CPFSK and $h = 1/2$, the coding rate is two thirds (three signals for two input bits), thus, the theoretical limits for 1Tx-1Rx is −5.26 dB, for 2Tx-1Rx is −7.65 dB, and finally for 2Tx-2Rx is −7.14 dB.

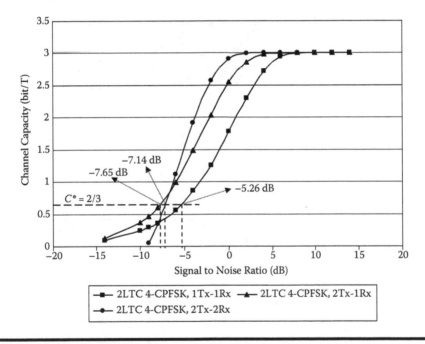

**Figure 11.10 Channel capacity $C^*$ of band-limited AWGN channels for 2LTC-4CPFSK $h = 1/2$ signals for 1Tx-1Rx, 2Tx-1Rx, and 2Tx-2Rx antenna configurations. (Copyright permission by John Wiley & Sons [15].)**

In the simulations, 2LTC-4CPFSK is considered and a one-third coding rate is used for each level. Therefore, the coding rate is two-thirds (three signals for two input bits). The bit error probability (Pb) versus signal energy-to-noise ratio (Es/N0) curves of 2LTC-4CPFSK is simulated for AWGN, Rician (for Rician channel parameter K = 10 dB) and Rayleigh channels for 1Tx-1Rx, 2Tx-1Rx, and 2Tx-2Rx antenna configurations. The theoretical limits of the 2LTC-4CPFSK scheme at $C^* = 2/3$ are obtained as −5.26 dB, −7.65 dB, and −7.14 dB for 1Tx-1Rx, 2Tx-1Rx, and 2Tx-2Rx antenna configurations, respectively, as in Figure 11.11. Perfect CSI, BW estimation, and no CSI cases are considered for fading channels and the results are shown from Figure 11.11 to Figure 11.17 for various frame sizes and antenna configurations.

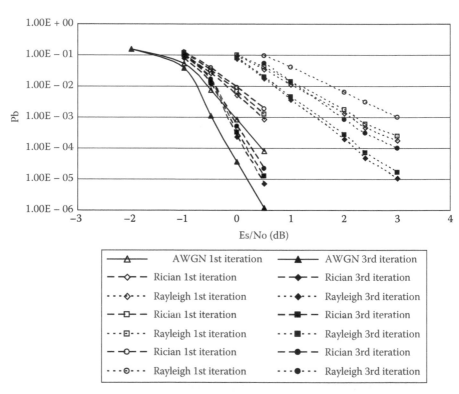

**Figure 11.11    Bit error probabilities of 1Tx-1Rx 2LTC-4CPFSK for *N* = 100 over AWGN, Rician (for K = 10 dB), and Rayleigh channels in perfect CSI, BW estimation, and no CSI cases. (Copyright permission by John Wiley & Sons [15].)**

In Figure 11.11, $1 \times 10^{-4}$ Pb at the 3rd iteration can be reached at –0.15 dB, 0.15 dB, and 2.3 dB for 1Tx-1Rx 2LTC-4CPFSK for AWGN, Rician, and Rayleigh channels in the case of perfect CSI, respectively. Bit error probabilities for 1Tx-1Rx, 2Tx-1Rx, and 2Tx-2Rx antenna configurations are shown in Figure 11.12, Figure 11.13, and Figure 11.14, respectively, for the frame size $N = 100$.

In Figure 11.12, for 2Tx-1Rx 2LTC-4CPFSK, $1 \times 10^{-4}$ Pb at the 3rd iteration can be reached at –1.1 dB, –0.65 dB, and 0.8 dB for AWGN, Rician, and Rayleigh channels, respectively, in the case of perfect CSI.

In Figure 11.13, for 2Tx-2Rx 2LTC-4CPFSK, $1 \times 10^{-4}$ Pb at the 3rd iteration can be reached at –4.2 dB, –3.7 dB, and –2.1 dB for AWGN, Rician, and Rayleigh channels, respectively, in the case of perfect CSI. For

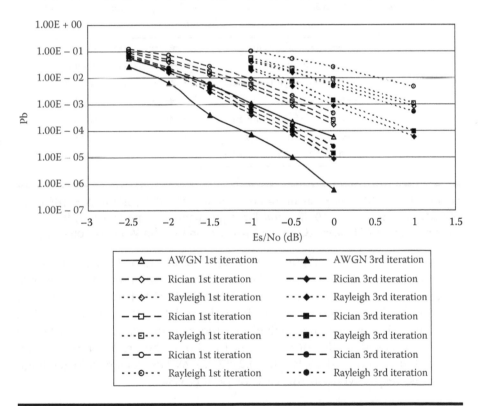

**Figure 11.12** **Bit error probabilities of 2Tx-1Rx 2LTC-4CPFSK for** $N$ = 100 **over AWGN, Rician (for K = 10 dB), and Rayleigh channels in the cases of perfect CSI, BW estimation, and no CSI cases. (Copyright permission by John Wiley & Sons [15].)**

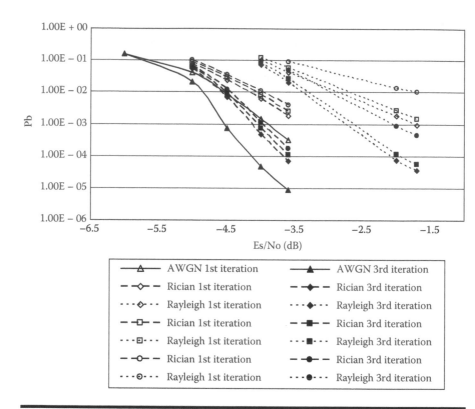

**Figure 11.13   Bit error probabilities of 2Tx-2Rx 2LTC-4CPFSK for *N* = 100 over AWGN, Rician (for K = 10 dB), and Rayleigh channels in perfect CSI, BW estimation, and no CSI cases. (Copyright permission by John Wiley & Sons [15].)**

the frame size *N* = 256, Figures 11.14, 11.15, and 11.16 show the bit error performance for 1Tx-1Rx, 2Tx-1Rx, and 2Tx-2Rx antenna configurations.

In Figure 11.14, for 1Tx-1Rx 2LTC-4CPFSK, $1 \times 10^{-4}$ Pb at the 3rd iteration can be reached at −0.45 dB, 0 dB, and 1.6 dB for AWGN, Rician, and Rayleigh channels, respectively, in the case of perfect CSI. In Figure 11.15, for 2Tx-1Rx 2LTC-4CPFSK, $1 \times 10^{-4}$ Pb at the 3rd iteration can be reached at −1.5 dB, −1.1 dB, and 0.2 dB for AWGN, Rician, and Rayleigh channels, respectively, in the case of perfect CSI.

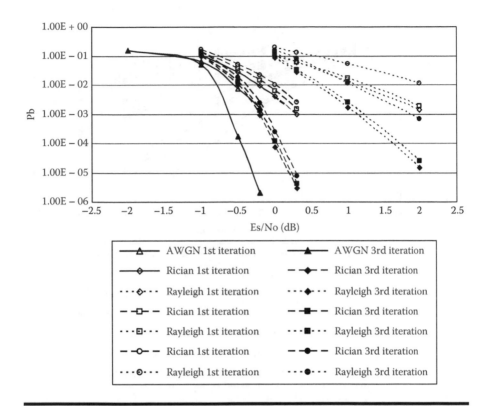

**Figure 11.14  Bit error probabilities of 1Tx-1Rx 2LTC-4CPFSK for $N$ = 256 over AWGN, Rician (for K = 10 dB), and Rayleigh channels in perfect CSI, BW estimation, and no CSI cases. (Copyright permission by John Wiley & Sons [15].)**

Finally, in Figure 11.16, for 2Tx-2Rx 2LTC-4CPFSK, $1 \times 10^{-4}$ Pb at the 3rd iteration can be reached at –4.5 dB, –4.1 dB, and –2.6 dB for AWGN, Rician, and Rayleigh channels, respectively, in the case of perfect CSI.

Bit error performance of the proposed coding technique with a BW channel parameter estimation is very close to the one using perfect CSI. Approximately 0.1 dB and 0.75 dB gains in Es/N0 are observed when the BW channel estimator is used, compared to no CSI case for Rician and Rayleigh channels, respectively.

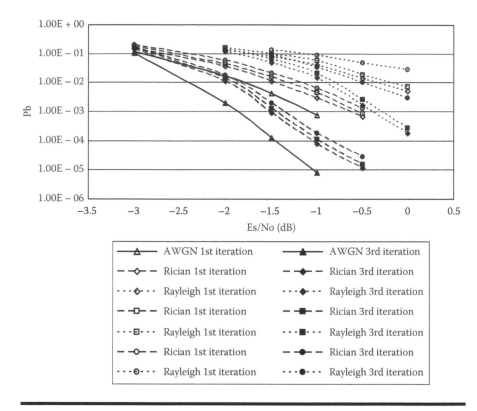

**Figure 11.15   Bit error probabilities of 2Tx-1Rx 2LTC-4CPFSK for *N* = 256 over AWGN, Rician (for K = 10 dB), and Rayleigh channels in perfect CSI, BW estimation, and no CSI cases. (Copyright permission by John Wiley & Sons [15].)**

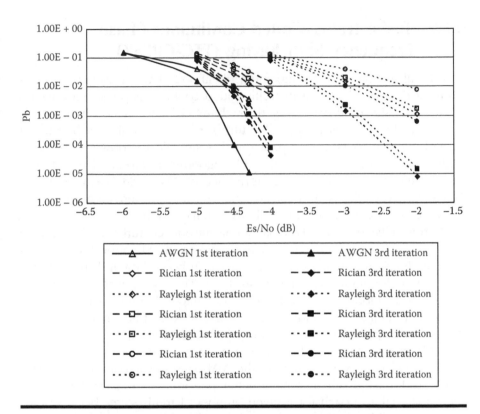

**Figure 11.16** **Bit error probabilities of 2Tx-2Rx 2LTC-4PFSK for** $N = 256$ **over AWGN, Rician (for K = 10 dB), and Rayleigh channels in perfect CSI, BW estimation, and no CSI cases. (Copyright permission by John Wiley & Sons [15].)**

## 11.2 Turbo Trellis–Coded Continuous-Phase Frequency Shift Keying (TTC-CPFSK)

In this subsection, a blind maximum likelihood channel estimation algorithm is developed for turbo trellis-coded/continuous-phase modulation (TTC/CPM) signals propagating through AWGN and Rician fading environments [17]. CPM is adapted for turbo trellis-coded signals because it provides a low spectral occupancy and is suitable for power and bandwidth-limited channels. Here, the BW algorithm is modified to estimate the channel parameters. The performance of TTC/CPM is simulated for 16-CPFSK over AWGN and Rician channels for different frame sizes, in the case of ideal CSI, no CSI, and BW estimated CSI.

Berrou, Glavieux and Thitimasjshima introduced turbo codes in 1993 [4], which achieved high performance close to the Shannon limit. Subsequently, Blackert and Wilson concatenated turbo coding and trellis-coded modulation for multilevel signals in 1996. The result was turbo trellis-coded modulation (TTCM) [21]. In band-limited channels, such as deep space and satellite communications, CPM has explicit advantages because it has a low spectral occupancy property. CPM is composed of CPE and MM. CPE is a convolutional encoder producing code word sequences that are mapped onto waveforms by the MM, creating a continuous phase signal. CPE-related schemes have better performance than systems using the traditional approach for a given number of trellis states because they increase Euclidean distance. Once the memory structure of CPM is assigned, it is possible to design a single-joint convolutional code, composed of trellis- and convolutionally-coded CPM systems as in [1–3].

To improve error performance and bandwidth efficiency, CPM and TTCM are combined creating turbo trellis-coded/continuous-phase modulation (TTC/CPM) [17]. The encoder includes a combined structure of convolutional encoder (CE) and CPE. Because providing the continuity of the signal for punctured codes is not possible when encoders are considered separately, the solution is to make a suitable connection between the encoders. In [17], a combined encoder structure called continuous-phase convolutional encoder (CPCE) is introduced and different dimensional CPFSK can be used. As the number of dimension increases, it influences the Euclidean distance and therefore the bit error ratio is improved.

In many cases, digital transmission is subject to inter symbol interference (ISI) caused by time dispersion of the multipath channels. In general, an adaptive channel estimator (i.e., based on the least mean square or recursive least square algorithms) is implemented in parallel to a trellis-based equalizer (e.g., maximum likelihood (ML) sequence estimator). This approach estimates channel parameters and input data jointly [22]. A further development is the introduction of the per-survivor-processing technique [23], in which each survivor employs its own channel estimator and no tentative decision delay is afforded. To exploit properties of multipath fading and track time varying channels, model-fitting algorithms are used in [24–26]. The ML joint data and channel estimator in the absence of training is presented in [27]. Although all of these trellis-based techniques are applied for coherent detection, noncoherent blind equalization techniques are an interesting alternative area of current research [28–30].

In this section, an efficient turbo trellis-coding structure is presented to generate the multidimensional continuous-phase signals. Then, a general ML blind method is investigated to estimate the Rician channel parameters for these signals. In [17], the authors concentrate on ML, because these estimation algorithms perform satisfactorily, even if there is only a short data record available [31]. The blind parameter is assumed to be constant in this process; however, implementation of the ML method directly for the blind problems results in a computational burden. The finite memory of the TTC/CPM along with the independent identical distribution (i.i.d.) and finite alphabet structure of the input data allow the system response to be modeled as a Markov chain [32]. However, the Markov chain is hidden because it can only be inferred from the noisy observation [33]. In [17], the hidden Markov model (HMM) formulation is considered and the computationally efficient BW algorithm is employed [32] in order to estimate Rician channel parameters.

## 11.2.1 Encoder Structure

The most critical point is to keep the continuity of the signals, in spite of the discontinuity property of the puncturer. In Equation (11.3), $\phi_0$ corresponds to the initial angle of the instant signal and the final angle of the previous signal. The final angle of the previous signal should be kept to provide continuity. If there is no puncturer block at the transmitter, the trellis states of joint structure indicate the final and initial angles. Because the transmitter of [17] includes a puncturer, previous information of the encoder is lost, thus continuity must be checked. This is because previous information of the considered branch cannot be used, since the puncturer switches to the other branch in the following coding step. Another drawback is a deinterleaver block at the output of the lower branch of the TTCM encoder as in [17].

First the deinterleaver is removed at the transmitter side and then the interleaver structure is changed in the decoder to reassemble the data. In the TTCM encoder, there are two parallel identical encoders, but only one of them is used instantly. The criterion behind choosing the signal is based on the next state of the other encoder's CPE. Thus, the authors in [16,17] use the inputs of the other encoder's CPE memory units as inputs of the MM in addition to the encoder outputs, so the continuity is maintained. Figure 11.17 shows the general structure of the TTC/CPM encoder. Any number of memory units can be used in CPE, but connections from the input of the memories to the MM are to be set and these result in the variation of the modulation index $h$ and signal space.

A concrete solution is studied for continuous-phase signaling by not using extra memories for CPE. Some of the CE memories for CPE are common and thus, the TTCM encoder structure complexity is kept constant. The structure that comprises CE and CPE is called continuous-phase convolutional encoder (CPCE). This combined structure is used not only to code the input data, but also to provide the continuity. As an example, an encoder structure and input-output table for 16-CPFSK are shown. Whereas there are two data streams as inputs. In Figure 11.18, the TTC/CPM structure is shown for 16-CPFSK. Here, input bit $m = 2$ and code rate $R = 1/2$. In this encoder structure, there is only one memory unit for CPE. Thus, modulation index $h$ is designated as $1/2$, although there is one additional input to MM. The last memory unit is also used as a CPE memory, since there is a close relation between the signal phase state and the CPE memory. If the memory unit value is 0, the signal phase will be 0 and if the memory is 1, the signal phase will be 1. If there is only one memory unit for CPE, there should be only two phase states and if are two phase states, $h = 1/2$. If two memory units are used for CPE, there are four phases (0, $\pi/2$, $\pi$, $3\pi/2$), then $h$ becomes $1/4$. But, if two memories are used, there are five outputs. Therefore, modulation must be 32-CPFSK for $h = 1/4$. This means there is no alternative for 16-CPFSK other than $h = 1/2$. Moreover, there is no alternative for 32-CPFSK other than $h = 1/4$ for the proposed system, while $m = 2$.

In Figure 11.18, at the upper encoder, CE1 consists of three memory units and CPE1 uses the last memory unit of CE1. At the lower encoder,

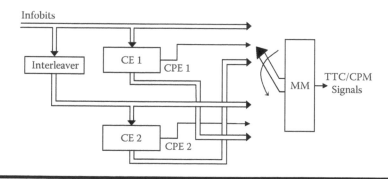

**Figure 11.17　TTC/CPM encoder structure for M-CPFSK with any $h$ value. (Copyright permission from IEEE [17].)**

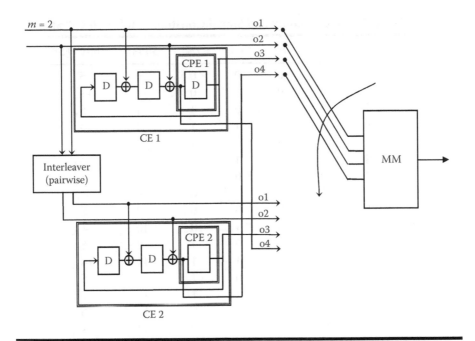

**Figure 11.18** **TTC/CPM encoder structure for 16-CPFSK and *h* = 1/2. (Copyright permission from IEEE [17].)**

CE2 consists of three memory units and CPE2 uses the last memory unit of CE2. When it is the upper encoder's turn, use CE1 of the upper and CPE2 of the lower encoder. This combined structure (CE1+CPE2) is defined as CPCE1. If it is the lower encoder's turn, use CE2 of the lower and CPE1 of the upper encoder. This combined structure (CE2+CPE1) is defined as CPCE2.

In Table 11.2, input–output data and signal constellations are given. Here, "o1" and "o2" systematic bits show the first two outputs and refer to the input bits "i1" and "i2," respectively; "o3" is the coded bit and "o4" is the input data of the third memory of the other encoder that is used for CPFSK; and "o4" indicates at which phase the signal will start at the next coding interval. If "o4" is 0, the end phase of the instant signal and the starting angle of the next signal phase is 0. If "o4" is 1, the end phase of the instant signal and the starting angle of the next signal phase is $\pi$. If "o3" is 0, the signal is positive and if "o3" is 1, the signal is negative. Only if these conditions are met is continuity granted. According to the Figure 11.19 and Table 11.2, the transmitted signals at the phase transitions are as follows: from 0 phase to 0 phase, $s_0$ $s_4$ $s_8$ $s_{12}$; from 0 phase to $\pi$ phase, $s_1$ $s_5$ $s_9$ $s_{13}$; from $\pi$ phase to 0 phase, $-s_2$ $-s_6$ $-s_{10}$ $-s_{14}$, and from $\pi$ phase to $\pi$ phase, $-s_3$ $-s_7$ $-s_{11}$ $-s_{15}$.

**Table 11.2   Input–Output and Signal Constellation for 16-CPFSK**

| i1 | i2 | o1 | o2 | o3 | o4 | Signal | 16-CPFSK |
|----|----|----|----|----|----|--------|----------|
| 0 | 0 | 0 | 0 | 0 | 0 | 0 | $S_0$ |
| 0 | 0 | 0 | 0 | 1 | 0 | 2 | $-s_2$ |
| 0 | 0 | 0 | 0 | 0 | 1 | 1 | $S_1$ |
| 0 | 0 | 0 | 0 | 1 | 1 | 3 | $-s_3$ |
| 0 | 1 | 0 | 1 | 0 | 1 | 5 | $S_5$ |
| 0 | 1 | 0 | 1 | 1 | 1 | 7 | $-s_7$ |
| 0 | 1 | 0 | 1 | 0 | 0 | 4 | $S_4$ |
| 0 | 1 | 0 | 1 | 1 | 0 | 6 | $-s_6$ |
| 1 | 0 | 1 | 0 | 0 | 0 | 8 | $S_8$ |
| 1 | 0 | 1 | 0 | 1 | 0 | 10 | $-s_{10}$ |
| 1 | 0 | 1 | 0 | 0 | 1 | 9 | $S_9$ |
| 1 | 0 | 1 | 0 | 1 | 1 | 11 | $-s_{11}$ |
| 1 | 1 | 1 | 1 | 0 | 1 | 13 | $S_{13}$ |
| 1 | 1 | 1 | 1 | 1 | 1 | 15 | $-s_{15}$ |
| 1 | 1 | 1 | 1 | 0 | 0 | 12 | $S_{12}$ |
| 1 | 1 | 1 | 1 | 1 | 0 | 14 | $-s_{14}$ |

*Source:* Osman O., Uçan O.N., *IEEE Transactions on Wireless Communications*, Vol. 4, No. 2, pp. 397–403, March 2005.

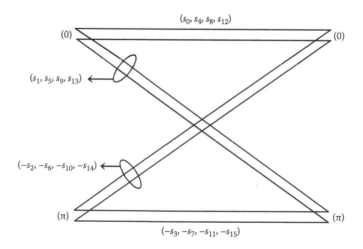

**Figure 11.19   16-CPFSK $h = 1/2$ signal phase diagram. (Copyright permission by IEEE [17].)**

## 11.2.2 *Channel Capacity for CPFSK and TTC/CPM*

It is very important to know the channel capacity of the system for different values of the CPFSK parameter, M. First, deal only with M-CPFSK. If there are M dimensions, the number of encoder outputs and the number of inputs of MM, O, can be calculated as $O = \log_2(M)$. In general, except for TTC-CPM, the signal constellation is consist of $V = 2M$ signals, but for TTC-CPM, $V = M$, as shown in Table 11.2; there are 16 different signals $\{s^0, s^1, \dots, s^{V-1}\}$ for 16-CPFSK. This decrease in the number of signal sets causes a slight decrease in channel capacity. The output of the AWGN channel is

$$y_n = s_n + w_n \tag{11.26}$$

where $s$ denotes a real valued discrete channel signal transmitted at modulation time $nT$, and $w$, is an independent normally distributed noise sample with zero mean and variance $\sigma^2$ along each dimension. The average SNR is defined as

$$SNR = \frac{E\left\{ \left| s_n^2 \right| \right\}}{E\left\{ \left| w_n^2 \right| \right\}} = \frac{E\left\{ \left| s_n^2 \right| \right\}}{M\sigma^2} \tag{11.27}$$

Normalized average signal power $E\left\{ \left| s_n^2 \right| \right\} = 1$ is assumed. Channel capacity can be calculated from the formula given in [20], if the codes with equiprobable occurrence of channel input signals are of interest.

$$C^* \approx \log_2(V) - \frac{1}{V} \cdot \sum_{k=0}^{V-1} E\left\{ \log_2 \sum_{i=0}^{V-1} \exp\left[ -\frac{\left| s^k + w - s^i \right|^2 - \left| w \right|^2}{2\sigma^2} \right] \right\} \tag{11.28}$$

in bit/T. Using a Gaussian random number generator, $C^*$ has been evaluated by Monte Carlo averaging of Equation (11.28). In Figure 11.20 channel capacity is plotted as a function of SNR for different dimensional CPFSK and TTC/CPM for 16-CPFSK. Here, it is obvious that the systems with CPFSK have quite high channel capacities compared to two-dimension and one-dimension related modulation schemes, such as MPSK, QAM, QPSK, BPSK, and so forth.

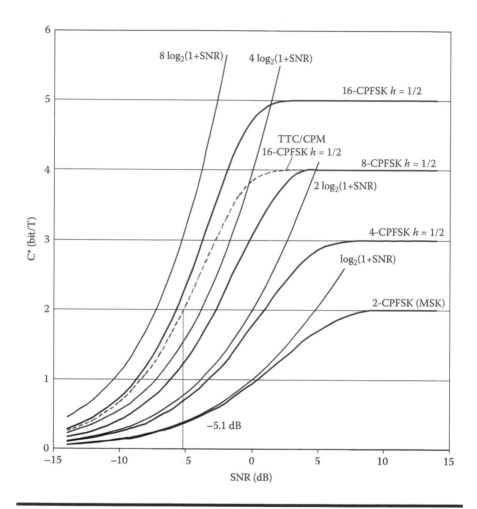

**Figure 11.20  Channel capacity $C^*$ of bandlimited AWGN channels for the TTC-CPM signal and different dimensional CPFSK signals. (Dotted line shows TTC-CPM for 16-CPFSK, $h = 1/2$. Bold lines show various M-CPFSK, and thin lines show various asymptotic curves.) (Copyright permission by IEEE [17].)**

In Figure 11.20, the dotted line shows the channel capacity for TTC/CPM in the case of 16-CPFSK $h = 1/2$. Maximum channel capacity is 4, because $V = M = 16$, and $C^*_{max} = \log_2(16)$. For the TTC/CPM of 16-CPFSK $h = 1/2$ and $m = 2$ for the transmission of 2 bit/T, theoretical limit is −5.1 dB. For the low channel capacities and with some error assumptions [44], channel capacity can be generalized according to the dimension $M$ of the modulation as below,

$$C^* \approx \frac{M}{2}\log_2(1 + SNR). \tag{11.29}$$

## 11.2.3  Decoder Structure

Here, in the turbo trellis decoder, the symbol-based decoder is equalized. Deinterleaver is not used in the encoder, thus symbol interleaver of the decoder only interleaves the odd indexed symbols that are created at the upper branch of the encoder. The even indexed symbols are interleaved in the encoder. In the encoder, the memory structure is kept constant, but the signal constellation is changed. In the decoding process, we use input data and parity data (o1 o2 o3) as mentioned in the example before, the "o4" is used only to provide continuity. There are different signals that have the same input and parity data. As an example, in Table 11.2, the first and third rows have the same "o1 o2 o3" but "o4" makes the first row "s0" and third row "s1," and "o4" has no role in the decoding process, because the inputs of the two CE are different, originating from the interleaver. The CPM demodulator finds the M dimensional signal from the M orthonormal functions for every observed signal. In metric calculation, the probability of every observed demodulated M dimensional signal $P(Y|X)$ is computed according to the Gaussian distribution. X and Y indicate transmitted and received data, respectively. In general, there are M different signals, some sharing the same inputs and parity data. The most probable one (MPO) chooses the most probable metric in the signal set, in which the m input bits and parity bit are the same. The formulation of the MPO is given by

$$\mathbf{MPO} = [\max(\mathbf{MPO_1}),\ \max(\mathbf{MPO_2}),\dots,\max(\mathbf{MPO_F})] \qquad (11.30)$$

**MPO** is a vector that contains $F = 2^{m+1}$ elements for every observed signal. $F$ is the number of the states of CE (here for $m = 2$, there is $m+1$ memories, thus $F = 8$), and $\mathbf{MPO_n}$ is a set of metrics that has the same "o1 o2 o3". The number of elements in $\mathbf{MPO_n}$ depends on the modulation index $h$ or the number of memories ($\eta$) in CPE. There are $P = 2^\eta$ elements in each $\mathbf{MPO_n}$. Therefore, **MPO** is obtained from $F \times P = 2^{m+1} \times 2^\eta = 2^{\eta+m+1}$ elements. (In our example $m = 2$, and $\eta = 1$, $F \times P = 2^{1+2+1} = 16 = V$).

At the decoder, the metric-s calculation is only used in the first decoding step. Because of the structure and modulation of the TTCM encoder, systematic bits and parity bits are mapped to one signal, and at the receiver, parity data cannot be separated from the received signal. Hence, in the first half of the first decoding for the S-b-S MAP decoder-1, *a priori* information hasn't been generated. Metric-s calculates the *a priori* information from the even index in the received sequence.

The next stage in the decoding process is an ordinary MAP algorithm. In Figure 11.21, an asterisk (*) shows the switch position when the even-indexed signals pass.

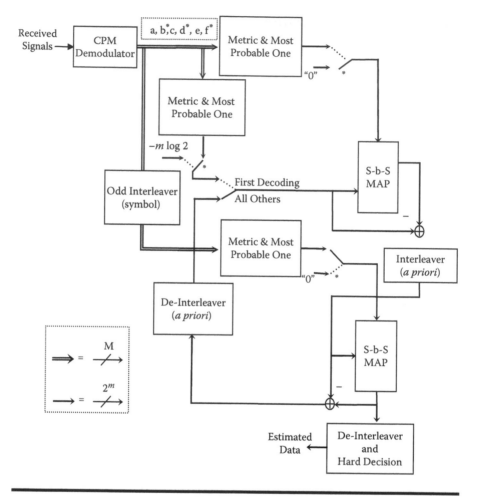

**Figure 11.21  TTC/CPM decoder structure. (Copyright permission by IEEE [17].)**

## 11.2.4  Blind Equalization

Consider the block diagram of the TTC/CPM scheme shown in Figure 11.17. At the $i$th epoch, the $m$-bit information vector $\mathbf{d}(i) = (d_1(i), d_2(i), \ldots, d_m(i))^T$ is fed to the TTC/CPM system. The output of this system is denoted as $g(\mathbf{d}(i))$. If the TTC/CPM signals are transmitted over a nondispersive, narrow-band fading channel, the received signal has the form of

$$y(i) = \rho(i)g(\mathbf{d}(i)) + n(i) \qquad (11.31)$$

where $x(i) = g(\mathbf{d}(i))$. The following assumptions are imposed on the received signal model:

> **AS1:** $n(i)$ is a zero mean, independent identical distributed (i.i.d.) AWGN sequence with the variance of $\sigma$.
> **AS2:** $\mathbf{d}(i)$ is an $m$ digit binary, equiprobable information sequence input to the TTC/CPM scheme.
> **AS3:** $\rho(i)$ represents a normalized fading magnitude having a Rician probability density function.
> **AS4:** Slowly varying nondispersive fading channels are considered.

The basic concern is the estimation of only the nondispersive Rician fading channel parameters from a set of observations. Due to its asymptotic efficiency, emphasis should be put on the ML approach [33] for the problem in hand. Based on the assumption AS1, the maximum likelihood metric for $\mathbf{d}$ is proportional to the conditional probability density function of the received data sequence of length $N$ (given $\mathbf{d}$) and is in the form of

$$m_l(y,\mathbf{d}) = \sum_{i=1}^{N} \sum_{z=1}^{O} |y(i) - \rho_l g_z(\mathbf{d}(i))|^2 \qquad (11.32)$$

where $\mathbf{d} = (d(1), d(2), \dots , d(N))$, $O$ is the number of signals that have the same "o1 o2 o3" output values and $g_z$ is a function of MM of the state of the CPCE and the value from the other encoder. As an example, in Figure 11.20, there is only one connection from the other encoder. Thus, $O$ is 2 and also there are 2 $g$ functions; first one is used when "o4" is zero and the second is used when "o4" is one. Since the transmitted data sequence is not available at the receiver, the ML metric to be optimized can be obtained by taking the average of $m_l(y,\mathbf{d})$ over all possible transmitted sequences $d$, which is expressed as

$$m(y) = \sum_{\mathbf{d}} m(y,\mathbf{d})$$

Then, the ML estimator of $\lambda = [\rho_l, \sigma_l^2]$ is the one that maximizes the average ML metric

$$\hat{\lambda} = \arg\ \max_{\lambda}[m_l(y)].$$

In this process, $\rho$ remains constant during the $l$ observation interval according to the **AS4**. The metric for **d** can be obtained by taking the average of Equation (11.32) over all possible values of $\mathbf{d}(i)$, which is formulated as

$$m_l(y) = \sum_d \sum_{i=1}^{N} \sum_{z=1}^{O} |y(i) - \rho_l g_z(\mathbf{d}(i))|^2 \qquad (11.33)$$

Because the direct minimization of Equation (11.33) requires the evaluation of Equation (11.32) over all possible transmitted sequences, the detection complexity increases exponentially with the data length $N$. Fortunately, a general iterative approach to the computation of ML estimation, BW algorithm, can be used to significantly reduce the computational burden of an exhaustive search. Although, only a narrow-band Rician channel is considered, equalizing the effects of partially response signaling is necessary.

### 11.2.4.1   Baum–Welch Algorithm

The BW algorithm is an iterative ML estimation procedure that was originally developed to estimate the parameters of a probabilistic function of a Markov chain [32]. The evaluation of the estimates in the iterative BW algorithm starts with an initial guess $\lambda^{(0)}$. At the $(t + 1)^{th}$ iteration, it takes estimate $\hat{\lambda}^{(t)}$ from the $t$th iteration as an initial value and re-estimates $\hat{\lambda}^{(t+1)}$ by maximizing $\Theta(\hat{\lambda}^{(t)}, \lambda')$ over $\lambda'$ [45].

$$\hat{\lambda}^{(t+1)} = \arg \min_{\lambda'} \Theta(\hat{\lambda}^{(t)}, \lambda') \qquad (11.34)$$

This procedure is repeated until the parameters converge to a stable point. It can be shown that the explicit form of the auxiliary function at the $i$th observation interval is in the form of

$$\Theta(\hat{\lambda}^{(t)}, \lambda') = C + \sum_{k=1}^{N} \sum_{m=1}^{F} \sum_{z=1}^{O} \gamma_m^{(t)}(i) \left\{ \frac{1}{\sigma_l'^2} \|y(i) - \rho_l' g_z(\phi_m)\|^2 \right\} \qquad (11.35)$$

where $\gamma_m^{(t)}(i) = f_{\lambda'^t}(y, \phi(i) = \phi_k)$ is the joint likelihood of data $y$ at state $\phi_k$ at time $i$ and $C$ is a constant. Using the definition of the backward $\beta_k(i)$ and the forward $\alpha_k(i)$ variables,

$$\beta_j^{(t)}(i) = \sum_{j=1}^{F} a_{m,j} \beta_j^{(t)}(i+1) f_j(y(i+1)) \qquad i = N-2, \ldots, 0 \qquad (11.36)$$

$$\alpha_j^{(t)}(i) = \sum_{m=1}^{F} a_{m,j} \alpha_j^{(t)}(i-1) f_j(y(i)) \qquad i = 1,\ldots,N-1 \qquad (11.37)$$

Here $a_{m,j}$ is the state transition probability of the CPCE and is always $1/F$. $f_j(y)$ is the probability density function of the observation and has a Gaussian nature,

$$f_k(y(i)) = \frac{1}{\sigma_l \sqrt{2\pi}} e \, \mathfrak{p} \left\{ -\frac{1}{2\sigma_l^2} \| y(i) - \eta_k \|^2 \right\} \qquad (11.38)$$

$\eta_k = \rho_l g_z(\phi_k)$ is the probable observed signal and $\sigma_l^2$ is the variance of the Gaussian noise. The initial conditions of $\beta$ and $\alpha$ are

$$\beta_j^{(t)}(N-1) = 1 \qquad 1 \le j \le F \qquad (11.39)$$

$$\alpha_m^{(t)}(0) = p_m(y(0)) \quad 1 \le m \le F \qquad (11.40)$$

$p_m$ is the initial probability of the signals and equals $1/F$. The probabilities of the partial observation sequences $\gamma_k(i)$, can be recursively obtained as [31–34], given the state $\phi_k$ at time $i$,

$$\gamma_k^{(t)}(i) = \alpha_k^{(t)}(i)\beta_k^{(t)}(i) \qquad (11.41)$$

The iterative estimation formulas can be obtained by setting the gradient of the auxiliary function to zero $\nabla_{\rho_i}\Theta = 0$, $\nabla_{\sigma_i^2}\Theta = 0$. Then the following formulas are applied until the stable point is reached.

$$\nabla_{\sigma_i^2}\Theta = \frac{N}{2\sigma_i'^2} + \frac{1}{2\sigma_i'^4}\sum_{i=l}^{N}\sum_{m=1}^{F}\sum_{z=1}^{O}\gamma_m^{(t)}(i)\| y(i) - \rho_j' g_z(\phi_m) \|^2 \quad (11.42)$$

$$\nabla_{\rho_i}\Theta = \frac{2}{\sigma_i'^2}\sum_{i=1}^{N}\sum_{m=1}^{F}\sum_{z=1}^{O}\gamma_m^{(t)}(i)\left\{ \left[ y(i) - \rho_j' g_z(\phi_m) \right] g_z(\phi_m) \right\} \quad (11.43)$$

## 11.2.5  Simulation Results of TTC/CPM for 16-CPFSK

In the following simulations, a two-input data sequence is taken and the corresponding TTC/CPM system for a 16-CPFSK is drawn as in Figure 11.18. The modulation index $h$ is chosen as one-half. The fading channel parameters are assumed to be constant during every 10 symbols and the Doppler frequency is not taken into account. In Figures 11.22 and 11.23, the bit error performances of TTC/CPM for 16-CPFSK are evaluated for the 1st and the 4th iterations while $N$ = 256 and 1024 respectively. The channel model is chosen as AWGN and Rician for $K$ = 4 dB and 10 dB in the case of ideal CSI, no CSI, and the BW channel estimation. It is interesting that the error probability of TTC/CPM is $1 \times 10^{-5}$ at –1.5 dB SNR value, but the similar BER

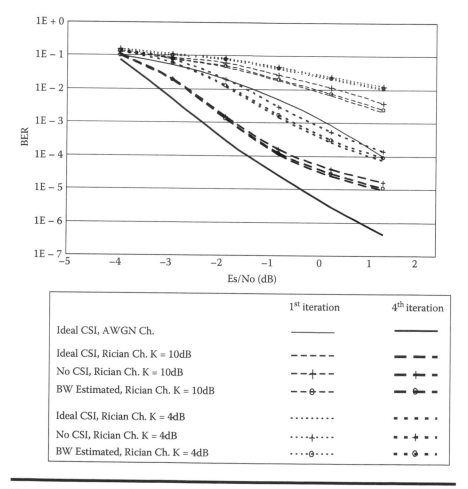

|  | 1st iteration | 4th iteration |
|---|---|---|
| Ideal CSI, AWGN Ch. | ——— | ——— |
| Ideal CSI, Rician Ch. K = 10dB | – – – – – | ▬ ▬ ▬ |
| No CSI, Rician Ch. K = 10dB | – –+– – | ▬ +▬ |
| BW Estimated, Rician Ch. K = 10dB | – –⊖– – | ▬ ⊖▬ |
| Ideal CSI, Rician Ch. K = 4dB | · · · · · · · · · | ▪ ▪ ▪ ▪ ▪ |
| No CSI, Rician Ch. K = 4dB | · · · ·+· · · · | ▪ ▪ +▪ ▪ · |
| BW Estimated, Rician Ch. K = 4dB | · · · ·⊖· · · · | ▪ ▪ ⊖▪ ▪ · |

**Figure 11.22  TTC/CPM for 16-CPFSK, $h$ = 1/2, $N$ = 256. (Copyright permission by IEEE [17].)**

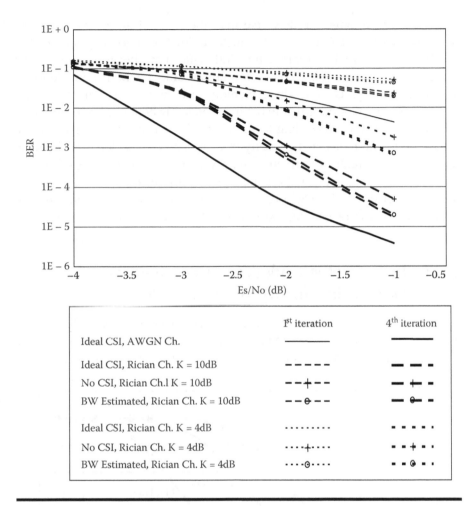

**Figure 11.23  TTC/CPM for 16-CPFSK, $h = 1/2$, $N = 1024$. (Copyright permission by IEEE [17].)**

is observed at 7.2 dB in TTCM for $N = 1024$ in AWGN channel as in [35]. This performance improvement also can be carried on for various channel models and different frame sizes. These results show the high performance of the proposed TTC/CPM compared with the classic schemes in various satellite channels. In the case of a BW channel parameter estimation, the bit error ratios of TTC/CPM are found to be very close to the ones in ideal cases.

In this subsection, the efficient coding method TTC/CPM is explained. Channel capacity of the proposed system and asymptotic curves of multidimensional modulations are given. According to the channel capacity curve of 16-CPFSK for TTC/CPM, error-free transmission of 2 bit/T theoretically possible at SNR = −5.1 dB. TTC/CPM simulation result for $N = 1024$ and

BER = $1 \times 10^{-5}$ is only 3.6 dB worse than the theoretical channel capacity. At the decoder, the BW channel parameter estimation algorithm is used. The combined structure has more bandwidth efficiency, because it has CPM characteristics. When using common memory units for both convolutional- and continuous-phase encoders, memory size and trellis state are kept constant. The combined structure has a considerably better error performance compared to TTCM. According to the channel capacity of M-CPFSK modulation, the multidimensional property of CPM can be considered as the main reason for such high performance improvement. We sketch the results for different SNRs, iteration numbers, frame sizes, and channel models. Finally, the results of the TTC/CPM, preprocessed with the BW algorithm and the results in the case of no CSI, are found to be extraordinary.

## 11.3 Turbo Trellis-Coded/Continuous-Phase Modulation for MIMO Channels

It is very important to know the channel capacity of the TTC/CPM for 1Tx-1Rx, 2Tx-1Rx, and 2Tx-2Rx antenna configurations. In Figure 11.24 there are four inputs to MM, therefore, the modulation is 16-CPFSK. If there are

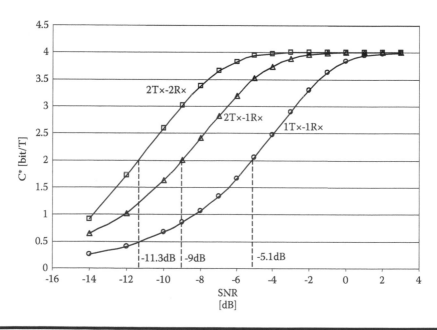

**Figure 11.24    Channel capacity C\* of bandlimited AWGN channels for 16-CPFSK, *h* = 1/2, TTC/CPM signals for 1Tx-1Rx, 2Tx-1Rx, and 2Tx-2Rx antenna configurations. (Copyright permission by IET [16].)**

M dimensions, the number of encoder outputs and the number of inputs of MM, $O$, can be calculated as $O = \log_2(M)$. In general, except for TTC/CPM, the signal constellation consists of $V = 2M$ signals, but for TTC/CPM, $V = M$ as shown in Table 11.2; there are 16 different signals $\{s^0, s^1, \ldots, s^{V-1}\}$ for 16-CPFSK. For 2Tx-1Rx configuration, transmitted signals from each antenna at the same time interval are the same; thus, the received signal is

$$y(n) = \rho_{11}(n)s(n) + \rho_{21}(n)s(n) + w(n)$$

$$= (\rho_{11}(n) + \rho_{21}(n))s(n) + w(n)$$

(11.44)

Here, $\rho_{11}$ and $\rho_{21}$ are fading amplitudes for each multipath channel. For 2Tx-2Rx configuration, the received signals at received antennas are as below:

$$y_1(n) - (\rho_{11}(n) + \rho_{21}(n))s(n) + w_1(n)$$

(11.45)

$$y_2(n) = (\rho_{12}(n) + \rho_{22}(n))s(n) + w_2(n)$$

(11.46)

Here, $y_1$ and $y_2$ are the received signals of the first and second receiver antennas. $\rho_{11}$, $\rho_{21}$, $\rho_{12}$, and $\rho_{22}$ are fading amplitudes for each multipath channel. $w_1$ and $w_2$ are noise values. The total received signal is

$$y(n) = (\rho_{11}(n) + \rho_{21}(n) + \rho_{12}(n) + \rho_{22}(n))s(n) + w_1(n) + w_2(n) \quad (11.47)$$

Channel capacities are computed for AWGN channels, therefore fading parameters are considered as 1. To calculate the channel capacity, Equation (11.28) is rewritten.

$$C^* \approx \log_2(V) - \frac{1}{V} \cdot \sum_{k=0}^{V-1} E\left\{ \log_2 \sum_{i=0}^{V-1} \exp\left[ -\frac{\left|s^k + w - s^i\right|^2 - \left|w\right|^2}{2\sigma^2} \right] \right\} \quad (11.48)$$

in bit/T. Using a Gaussian random number generator, $C^*$ has been evaluated by Monte Carlo averaging of Equation (11.48). In Figure 11.24, channel capacity is plotted as a function of SNR for different dimensional CPFSK and TTC/CPM for 16-CPFSK.

In Figure 11.24, the line with circles (O) shows the channel capacity for TTC/CPM in the case of 16-CPFSK $h = 1/2$. The maximum channel capacity is 4, because $V = M = 16$ and $C^*_{max} = \log_2(16)$. In our case, TTC/CPM for 16-CPFSK $h = 1/2$ and $m = 2$ for the transmission of 2 bit/T, theoretical limit is –5.1 dB. For the 2Tx-1Rx antenna configuration, channel capacity is shown with the line with triangles ($\triangle$), the transmission of 2 bit/T, theoretical limit is –9dB. Finally, for the 2Tx-2Rx antenna configuration, channel capacity is shown with the line with squares ($\square$), the transmission of 2 bit/T, theoretical limit is –11.3dB

In simulations, 1Tx-1Rx, 2Tx-1Rx, and 2Tx-2Rx antenna configurations are considered to show the performance of TTC/CPM over MIMO channels; 256 and 1024 frame sizes with 2 bit inputs are chosen. AWGN, Rician for $K = 10$ dB and $K = 4$ dB fading channels are taken into account and fading amplitudes are assumed to be known. Bit error probability of 4th iteration of TTC/CPM for 16-CPFSK and $h = 1/2$ is shown in Figure 11.25 for the frame size $N = 256$, and in Figure 11.26 for the frame size $N = 1024$. Circles, triangles, and squares are the pointers that indicate the antenna configurations, 1Tx-1Rx, 2Tx-1Rx, and 2Tx-2Rx respectively. Solid lines, dashed lines, and dotted lines show the channels as AWGN, Rician for $K = 10$ dB, and Rician for $K = 4$dB, respectively. In Figure 11.25 and Figure 11.26, bit error performances improve very fast as the number of

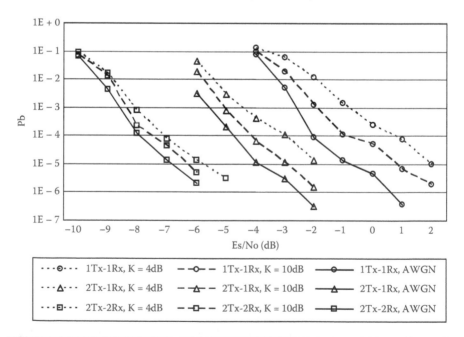

**Figure 11.25  Bit error probability of 4th iteration of TTC/CPM for 16-CPFSK, $h = 1/2$, $N = 256$ while 1Tx-1Rx, 2Tx-1Rx, and 2Tx-2Rx antenna configurations are considered. (Copyright permission by IET [16].)**

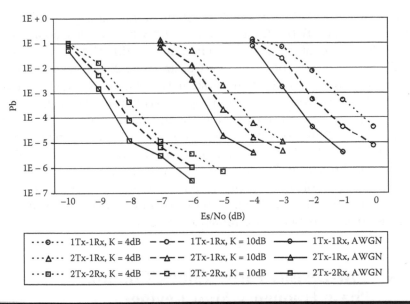

**Figure 11.26   Bit error probability of 4th iteration of TTC/CPM for 16-CPFSK, $h = 1/2$, $N = 1024$ while 1Tx-1Rx, 2Tx-1Rx, and 2Tx-2Rx antenna configurations are considered. (Copyright permission by IET [16].)**

antennas is increased. Bit error probability of $10^{-5}$ can be reached at $-1.9$ dB by 2Tx-1Rx and $-5.8$ dB by 2Tx-2Rx configurations while 1Tx-1Rx configuration is reached at 2 dB for Rician $K = 4$ dB fading channel. There are 3.9 dB and 7.8 dB gains because of using 2Tx-1Rx and 2Tx-2Rx instead of 1Tx-1Rx. For Rician $K = 10$ dB fading channel, bit error probability of $10^{-5}$ can be reached at $-2.9$ dB by 2Tx-1Rx and $-6.3$ dB by 2Tx-2Rx configurations while 1Tx-1Rx configuration is reached at 0.8 dB. There are 3.7 dB and 7.1 dB gains because of using 2Tx-1Rx and 2Tx-2Rx instead of 1Tx-1Rx. For the AWGN channel, a bit error probability of $10^{-5}$ can be reached at $-3.9$ dB by 2Tx-1Rx and $-6.8$ dB by 2Tx-2Rx configurations while 1Tx-1Rx configuration is reached at $-0.9$ dB. There are 3 dB and 5.9 dB gains because of using 2Tx-1Rx and 2Tx-2Rx instead of 1Tx-1Rx.

In Figure 11.26, a bit error probability of $10^{-5}$ can be reached at $-3$ dB by 2Tx-1Rx and $-6.9$ dB by 2Tx-2Rx configurations while 1Tx-1Rx configuration is reached at 1 dB for Rician $K = 4$ dB fading channel. There are 4 dB and 7.9 dB gains because of using 2Tx-1Rx and 2Tx-2Rx instead of 1Tx-1Rx. For Rician $K = 10$ dB fading channel, bit error probability of $10^{-5}$ can be reached at $-3.8$ dB by 2Tx-1Rx and $-7.2$ dB by 2Tx-2Rx configurations while 1Tx-1Rx configuration is reached at 0 dB. There are 3.8 dB and 7.2 dB gains because of using 2Tx-1Rx and 2Tx-2Rx instead of 1Tx-1Rx. For AWGN channel, a bit error probability of $10^{-5}$ can be reached at $-4.6$ dB by 2Tx-1Rx and $-7.9$ dB by 2Tx-2Rx configurations while 1Tx-1Rx configuration is reached at $-1.4$ dB. There are 3.2 dB and 6.5 dB gains because of using 2Tx-1Rx and 2Tx-2Rx instead of 1Tx-1Rx.

The channel capacity of the system is obtained for various antenna configurations. Diversity gains of using 2Tx-1Rx instead of 1Tx-1Rx at $10^{-5}$ bit error probability of 4th iteration for the frame size of 256 are 3 dB, 3.7 dB, and 3.9 dB for AWGN, Rician $K = 10$ dB, and Rician $K = 4$ dB, respectively. If the 2Tx-2Rx antenna configuration is used, the diversity gain would be 5.9 dB, 7.1 dB, and 7.8 dB. Moreover, if the frame size is chosen as 1024, diversity gains would be 3.2 dB, 3.8 dB, and 4 dB for 2Tx-1Rx and 6.5 dB, 7.2 dB, and 7.9 dB for 2Tx-2Rx. TTC/CPM has excellent bit error probabilities for MIMO channels.

## 11.4 Low-Density Parity Check/Continuous-Phase Frequency Shift Keying

To view the error performance of a digital communication system in which LDPC codes are used as a channel encoder and CPM is used as a modulator, a scheme is designed in [19], called low density parity check/continuous phase frequency shift keying (LDPC/CPFSK). The LDPCC-CPFSK system model is shown in Figure 11.27. As this figure shows, message bits are first coded by LDPC encoder. Then the encoded bits are turned into serial to $\log_2 M$ parallel branch according to the type of M-ary CPFSK modulation. The CPE encodes the least significant bit (LSB) again for phase continuity.

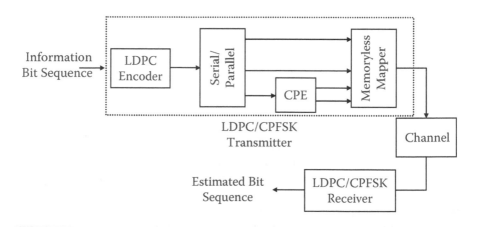

**Figure 11.27  LDPCC-CPFSK scheme. (Copyright permission by John Wiley & Sons [19].)**

The coded bits are mapped into M-ary CPFSK signals $s_i$, $i \in (1, 2, \ldots, 2M)$ by MM according to the partitioning rule.

In the partitioning rule, the most significant bit (MSB) of Serial to Parallel ($S/P$) converter output, $x_1$, divides the signal set into two subsets. If $x_1 = 0$, then the first subset is chosen; if $x_1 = 1$, then the second subset is chosen. Likewise the $x_2$ bit divides the subsets into two groups. This partitioning process continues until the last partitioning level occurs.

In order to clarify the operation of the partitioning mechanism for LDPCC-CPFSK system, the set partitioning for 8-ary CPFSK is chosen, as shown in Figure 11.28. Here, if $x_1 = 0$, then first subset $\{s_0, s_1, s_2, s_3, s_8, s_9, s_{10}, s_{11}\}$, if $x_1 = 1$, then the second subset $\{s_4, s_5, s_6, s_7, s_{12}, s_{13}, s_{14}, s_{15}\}$ is chosen. At the second partitioning level, we assume that the first subset is chosen in first partitioning level; $x_2$ determines whether $\{s_0, s_1, s_8, s_9\}$ or $\{s_2, s_3, s_{10}, s_{11}\}$ subsets are to be chosen. Similarly, in the first partitioning level, if the second

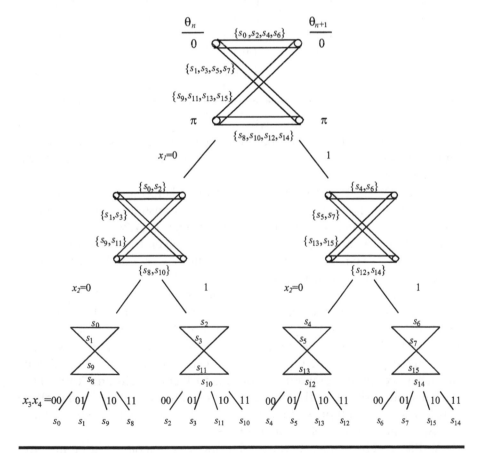

**Figure 11.28  8-CPFSK $h = 1/2$ signal phase diagram. (Copyright permission by John Wiley & Sons [19].)**

**Table 11.3   Input–Output and Signal Constellation for 8-CPFSK (Copyright permission by John Wiley & Sons [19]).**

| $\theta_N$ | $\beta_N$ | x1x2x3x4 | $\theta_{N+1}$ | 8-ary CPFSK |
|---|---|---|---|---|
| 0 | 0 | 0 0 0 0 | 0 | $s_0$ |
| 0 | 1 | 0 0 1 0 | $\pi$ | $s_1$ |
| 0 | 2 | 0 1 0 0 | 0 | $s_2$ |
| 0 | 3 | 0 1 1 0 | $\pi$ | $s_3$ |
| 0 | 4 | 1 0 0 0 | 0 | $s_4$ |
| 0 | 5 | 1 0 1 0 | $\pi$ | $s_5$ |
| 0 | 6 | 1 1 0 0 | 0 | $s_6$ |
| 0 | 7 | 1 1 1 0 | $\pi$ | $s_7$ |
| $\pi$ | 0 | 0 0 0 1 | $\pi$ | $s_8 = -s_0$ |
| $\pi$ | 1 | 0 0 1 1 | 0 | $s_9 = -s_1$ |
| $\pi$ | 2 | 0 1 0 1 | $\pi$ | $s_{10} = -s_2$ |
| $\pi$ | 3 | 0 1 1 1 | 0 | $s_{11} = -s_3$ |
| $\pi$ | 4 | 1 0 0 1 | $\pi$ | $s_{12} = -s_4$ |
| $\pi$ | 5 | 1 0 1 1 | 0 | $s_{13} = -s_5$ |
| $\pi$ | 6 | 1 1 0 1 | $\pi$ | $s_{14} = -s_6$ |
| $\pi$ | 7 | 1 1 1 1 | 0 | $s_{15} = -s_7$ |

subset is chosen, $x_2$ determines whether $\{s_4, s_5, s_{12}, s_{13}\}$ or $\{s_5, s_6, s_{14}, s_{15}\}$ subsets are to be chosen. In the last partitioning level, $x_3 x_4$ bits specify which signal will be transmitted to the channel. This procedure can be shown in Table 11.3.

In the LDPCC 8-ary CPFSK system, if modulation index ($h$) is chosen as one-half, initial and the ending phase of the transmitted signal will take two different values (0, $\pi$) as shown in Figure 11.28. In Table 11.3, input–output data and signal constellations are given for this system. Here, $x_1$ and $x_2$ are systematic bits, $x_3$ and $x_4$ are the bits encoded by CPE. Here, $x_3$ helps to show at which phase the signal will start at the next coding interval. If the initial phase ($\theta_n$) is "0" and "$x_3$" is 0, the ending phase of the instant signal and the starting angle of the next signal phase ($\theta_{n+1}$) is "0". If $\theta_n = 0$ and $x_3 = 1$, $\theta_{n+1} = \pi$. If $\theta_n = 1$ and $x_3 = 0$, $\theta_{n+1} = \pi$. If $\theta_n = 1$ and $x_3 = 1$, $\theta_{n+1}$ is 0. If the initial phase is "0", the signal is positive and, if the initial phase is "1", the signal is negative. Only if these conditions are met is continuity granted. According to Figure 11.28 and Table 11.4, the transmitted signals at the phase transitions are as follows: from 0 phase to 0, $s_0, s_2, s_4, s_6$; from 0 phase to $\pi$ phase, $s_1, s_3, s_5, s_7$; from $\pi$ phase to 0 phase, $s_9, s_{11}, s_{13}, s_{15}$ and from $\pi$ phase to $\pi$ phase, $s_8, s_{10}, s_{12}, s_{14}$.

At the receiver side, because the message passing algorithm—the decoding procedure in the LDPC decoder—uses one and zero probabilities

**Table 11.4   Degree Distribution of Irregular LDPC Code (Copyright permission by John Wiley & Sons [19]).**

| $x,y$ | $\lambda_x$ | $\lambda_y$ |
|---|---|---|
| 2 | 0.47718 | — |
| 3 | 0.28075 | — |
| 5 | 0.097222 | — |
| 7 | 0.0089286 | — |
| 8 | — | 1 |
| 14 | 0.00099206 | — |
| 15 | 0.1002 | — |

of received signal, for every decoding interval, one and zero probabilities are first evaluated for all parallel input branch sequences according to the partitioning rule. These probabilities are calculated as

$$P_0^L = \frac{1}{M}\left[\sum_{i=0}^{M-1}\frac{1}{\left(r_k - s_{2i}\right)^2}\right] \tag{11.49}$$

$$P_1^L = \frac{1}{M}\left[\sum_{i=0}^{M-1}\frac{1}{\left(r_k - s_{2i+1}\right)^2}\right] \tag{11.50}$$

where $r_k$ is received signal, $L$ is partitioning level and $s_i$ is the transmitted M-ary CPFSK signal.

After computing the one and zero probabilities as in (11.49) and (11.50), the received signal is mapped to a one-dimensional BPSK signal as follows,

$$\gamma^L = 1 - \frac{2 \cdot P_0^L}{P_0^L + P_1^L} \tag{11.51}$$

These probability computations and mapping are executed in every partitioning level according to the signal set, shown in Figure 11.28. For partitioning level 1–3, these probabilities are calculated as in Equations (11.52)–(11.57) where $P_i^L$, $i$ denotes zero or one probability, and $L$ denotes the partitioning level; $r_k$ is the received noisy CPFSK signal.

$$P_0^1 = \frac{1}{8}\left[\sum_{i=0}^{7}\frac{1}{\left(r_k - s_{2i}\right)^2}\right] \tag{11.52}$$

$$P_1^1 = \frac{1}{8}\left[\sum_{i=0}^{7}\frac{1}{\left(r_k - s_{2i+1}\right)^2}\right] \tag{11.53}$$

$$P_0^2 = \frac{1}{8}\left[\sum_{i=0}^{1}\left\{\frac{1}{\left(r_k - s_i\right)^2} + \frac{1}{\left(r_k - s_{i+4}\right)^2} + \frac{1}{\left(r_k - s_{i+8}\right)^2} + \frac{1}{\left(r_k - s_{i+12}\right)^2}\right\}\right] \tag{11.54}$$

$$P_1^2 = \frac{1}{8}\left[\sum_{i=0}^{1}\left\{\frac{1}{\left(r_k - s_{i+2}\right)^2} + \frac{1}{\left(r_k - s_{i+6}\right)^2} + \frac{1}{\left(r_k - s_{i+10}\right)^2} + \frac{1}{\left(r_k - s_{i+14}\right)^2}\right\}\right] \tag{11.55}$$

$$P_0^3 = \frac{1}{8}\left[\sum_{i=0}^{3}\left\{\frac{1}{\left(r_k - s_i\right)^2} + \frac{1}{\left(r_k - s_{i+8}\right)^2}\right\}\right] \tag{11.56}$$

$$P_1^3 = \frac{1}{8}\left[\sum_{i=0}^{3}\left\{\frac{1}{\left(r_k - s_{i+4}\right)^2} + \frac{1}{\left(r_k - s_{i+12}\right)^2}\right\}\right] \tag{11.57}$$

For every partitioning level, these one and zero probabilities are evaluated and the message bits estimated from Equations (10.8)–(10.12).

## 11.5  Simulations Results

In our study, we combine LDPC codes and the CPM structure together. For simulation purposes, we use (3, 6) regular and irregular that has a degree distribution shown in Table 11.4, LDPC codes. The block length is chosen at 504 bits and the number of maximum iteration is 100. In both regular and irregular codes, we design three LDPC coded CPFSK schemes for 4-ary CPFSK, 8-ary CPFSK and 16-ary CPFSK. The BER versus SNR curves of the proposed systems (LDPCC 4-ary CPFSK, LDPCC 8-ary CPFSK, LDPCC 16-ary CPFSK) are obtained for AWGN, Rician (for Rician channel parameter $K = 10$ dB) and Rayleigh channels through computer simulations. To compare the performance of these schemes, the BPSK system, that is encoded by the same LDPC codes, is also simulated. The results are shown in Figures 11.29–11.34. Although LDPCC 8-ary CPFSK, LDPCC 16-ary CPFSK systems have a higher bit/symbol rate than an LDPC-coded BPSK scheme, our proposed systems have better error performance in all channels and SNR values.

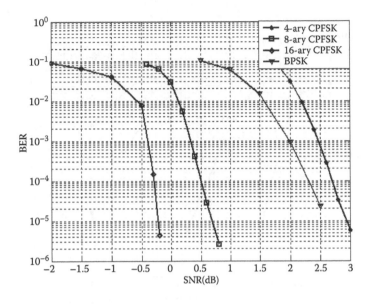

**Figure 11.29** Error performance of regular LDPCC-CPFSK system on AWGN channel. (Copyright permission by John Wiley & Sons [19].)

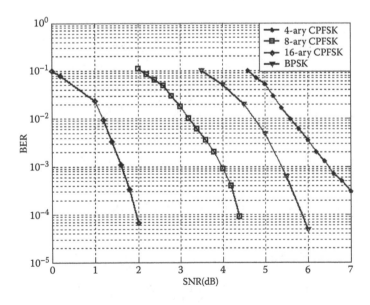

**Figure 11.30** Error performance of regular LDPCC-CPFSK system on Raleigh channel. (Copyright permission by John Wiley & Sons [19].)

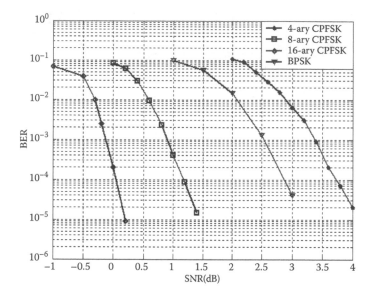

**Figure 11.31   Error performance of regular LDPCC-CPFSK system on Rician ($K = 10$ dB) channel. (Copyright permission by John Wiley & Sons [19].)**

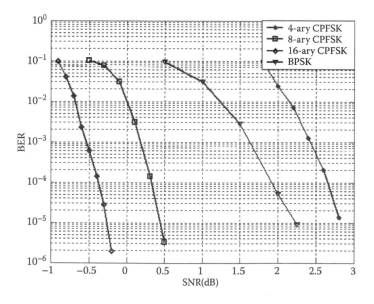

**Figure 11.32   Error performance of irregular LDPCC-CPFSK system on AWGN channel. (Copyright permission by John Wiley & Sons [19].)**

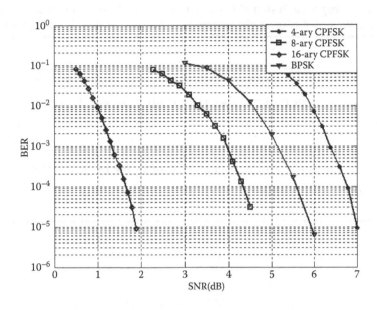

**Figure 11.33  Error performance of irregular LDPCC-CPFSK system on Raleigh channel. (Copyright permission by John Wiley & Sons [19].)**

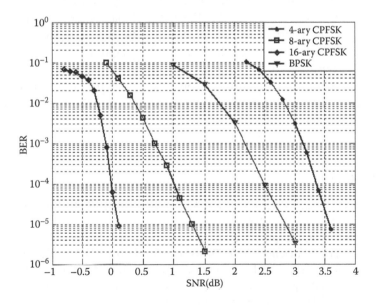

**Figure 11.34  Error performance of irregular LDPCC-CPFSK system on Rician ($K = 10$ dB) channel. (Copyright permission by John Wiley & Sons [19].)**

**Table 11.5    Proposed Systems' Coding Gains (in dB) over LDPCC-BSPK System for $P_b = 10^{-4}$ (Copyright permission by John Wiley & Sons [19]).**

| System | AWGN Channel | Rician Channel (K = 10 dB) | Rayleigh Channel |
|---|---|---|---|
| **Coding Gains for LDPCC-CPFSK over LDPCC-BPSK for Regular LDPC Code** | | | |
| 8-ary CPFSK | 1.78 | 1.62 | 1.48 |
| 16-ary CPFSK | 2.57 | 2.71 | 3.88 |
| **Coding Gains for LDPCC-CPFSK over LDPCC-BPSK for Irregular LDPC Code** | | | |
| 8-ary CPFSK | 1.59 | 1.5 | 1.3 |
| 16-ary CPFSK | 2.24 | 2.5 | 4.0 |

For a regular LDPC code, the proposed LDPCC-CPFSK systems provide coding gains of 1.48 dB up to 3.88 dB on various channels. Consequently irregular LDPCC-CPFSK systems have gains between 1.3 and 4 dB as seen in Table 11.5. As an example, LDPCC 8-ary CPFSK, LDPCC 16-ary CPFSK have 1.78 dB and 2.57 dB coding gains, respectively, on the AWGN channel for a bit error rate of $10^{-4}$ compared to the LDPCC-BPSK system. As seen in simulation results, LDPCC-BPSK has better error performance than LDPCC-4CPFSK, because BPSK has a double bit/symbol rate (1/2 bit/symbol) compared to LDPCC 4-ary CPFSK system (1 bit/symbol). Even though, LDPCC 8-ary CPFSK and 16-ary CPFSK systems also have a higher bit/symbol rate (1.5 and 2 bit/symbol) than LDPCC-BPSK system, they have significant coding gains because channel signals have 8 and 16 dimensions for 8-ary CPFSK and 16-ary CPFSK, respectively. Thus, as the modulated signal dimension (M) of M-ary CPFSK modulation increases, better error performances are obtained.

# References

[1] Rimoldi, B. E., A Decomposition Approach to CPM. *IEEE Trans. on Inform. Theory* 34:260–270, Mar. 1988.

[2] Rimoldi, B. E. Design of Coded CPFSK Modulation Systems for Band-Width and Energy Efficiency. *IEEE Trans. on Commun.* 37:897–905, Sept. 1989.

[3] Yang, R. H. H., and D. P. Taylor. Trellis-Coded Continuous-Phase Frequency-Shift Keying with Ring Convolutional Codes. *IEEE Trans. on Inform. Theory* 40:1057–1067, July 1994.

[4] Berrou, C., A Glavieux, and P. Thitimasjshima. Near Shannon-Limit Error Correcting Coding and Decoding: Turbo Codes (1). in Proc., *IEEE Int. Conf. on Commun.*, (Geneva, Switzerland):1064–1070, May 1993.

[5] Osman, O., and O. N. U•an. Blind Equalization of Turbo Trellis Coded Partial Response Continuous Phase Modulation Signaling over Narrow-Band Rician Fading Channels. *IEEE Trans. on Wireless Commun.,* Vol.4(2):397–403, Mar. 2005.

[6] U•an, O. N. Jitter and Error Performance of Trellis Coded Systems over Rician Fading Channels with Imperfect Phase Reference. *Electronic Letter* 32:1164–1166, 1996.

[7] Imai, H. and S. Hirakawa. A New Multilevel Coding Method Using Error-Correcting Codes. *IEEE Trans. On Inform. Theory* IT-23:371–377, May 1977.

[8] Yamaguchi, K., and H. Imai. Highly Reliable Multilevel Channel Coding System using Binary Convolutional Codes. *Electron. Letter* 23:939–941, Aug. 1987.

[9] Pottie G.J ., and D. P.Taylor. Multilevel Codes based on Partitioning. *IEEE Trans. Inform. Theory* 35:87–98, Jan. 1989.

[10] Calderbank, A. R. Multilevel Codes and Multistage Decoding. *IEEE Trans. on Commun.* 37:222–229, Mar. 1989.

[11] Forney, G. D. A Bounded-Distance Decoding Algorithm for the Leech Lattice with Generalization. *IEEE Trans. on Inform. Theory* 35, July 1989.

[12] Herzberg, H. Multilevel Turbo coding with Short Interleavers. *IEEE Journal on Selected Areas In Commun.*,16:303–309, Feb. 1998.

[13] Waschman, U., R. F. H. Fischer, and J. B. Huber. Multilevel Codes: Theoretical Concepts and Practical Design Rules. *IEEE Trans. on Inform. Theory* 45:1361–1391, July 1999.

[14] Osman, O., O. N. U•an, and N. Odabasioglu. Performance of Multi Level-Turbo Codes with Group Partitioning over Satellite Channels. *IEE Proc. Commun.* 152:1055–1059, Dec. 2005.

[15] Osman, O. Blind Equalization of Multilevel Turbo Coded-Continuous Phase Frequency Shift Keying over MIMO Channels. *IEE* 20(1):103–119, 2007

[16] Osman, O. Performance of Turbo Trellis Coded/Continuous Phase Modulation over MIMO Channels. *IET Proc. Communications* 1(3):354–358, 2007.

[17] Osman, O., and O. N. U•an O.N. Blind Equalization of Turbo Trellis-Coded Partial-Response Continuous-Phase Modulation Signaling over Narrow-Band Rician Fading Channels. *IEEE Transactions on Wireless Communications* 4(2):397–403, Mar. 2005.

[18] U•an, O. N., and O. Osman O. Blind Equalization of Space Time-Turbo Trellis Coded/Continuous Phase Modulation over Rician Fading Channels. *European Transactions on Telecommunications* 15(5)447–458, Sept.–Oct, 2004.

[19] Hekim, Y, N. Odabasioglu, O. N. U•an. Performance of Low Density Parity Check Coded Continuous Phase Frequency Shift Keying (LDPCC-CPFSK) over Fading Channels. *International Journal of Communication Systems,* published online, 30 Jun 2006, DOI: 10.1002/dac.824.

[20] Naraghi-Pour, M. Trellis Codes for 4-ary Continuous Phase Frequency Shift Keying. *IEEE Trans. on Commun.*,41:1582–1587, Nov. 1993.

[21] Blackert, W., and S. Wilson. Turbo Trellis Coded Modulation. in *Proc. CISS'96* Mar. 1996.

[22] Proakis, J. G. *Digital Communications,* 3rd ed. New York: McGraw-Hill, 1995.

[23] Raheli, R., A. Polydoros, and C. K. Txou. PSP: A General Approach to MLSE in Uncertain Environments. *IEEE Trans. Communications* COM43:354–364, Feb./Mar./Apr. 1995.

[24] Dai, Q., and E. Shwedyk. Detection of Bandlimited Signals over Frequency Selective Rayleigh Fading Channels. *IEEE Trans. Communications* 42:941–950, Feb./Mar./Apr. 1994.

[25] Zhang, Y., M. P. Fitz, and S. B. Gelfand. Soft Output Demodulation on Frequency-Selective Rayleigh Fading Channels Using AR Models. in *Proc. IEEE GLOBCOM'97* 1:327–330, Nov. 1997.

[26] Davis, L. M., and I. B. Collings. DPSK versus Pilot-Aided PSK MAP Equalization for Fast-Fading Channels. *IEEE Trans. Communications* 49:226–228, Feb. 2001.

[27] N. Seshadri. Joint Data and Channel Estimation using Blind Trellis Search Techniques. *IEEE Trans. Communications* 42:1000–1011, Feb. 1994.

[28] Yu, X., and S. Pasupathy. Innovations-based MLSE for Rayleigh Fading Channels. *IEEE Trans. Communications* 43:1534–1544, Nov. 1995.

[29] Fred, A. M., and S. Pasupathy. Nonlinear Equalization of Multi-Path Fading Channels with Non-coherent Demodulation. *IEEE JSAC* 14:512–520, Apr. 1996.

[30] Colavolpe, G., and R. Raheli. Noncoherent Sequence Detection. *IEEE Trans.Communications* 47:1376–1385, Sept. 1999.

[31] Erkurt, M., and J. G. Proakis. Joint Data Detection and Channel Estimation for Rapidly Fading Channels. In *Proc. IEEE Globecom*:910–914, Dec. 1992.

[32] Baum, L. E., T. Petrie, G. Soules, and N. Weiss. A Maximization Technique occurring in the Statistical Analysis of Probabilistic Functions of Markov Chains. *The Annals of Mathematical Statistics* 41:164–171, 1970.

[33] Rabiner, L. R. A Tutorial on Hidden Markov Models and Selected Applications in Speech Recognition. in *Proceeding of the IEEE* 77(2):257–274, Feb. 1989.

[34] Kaleh, G. K., and R. Vallet. Joint Parameter Estimation and Symbol Detection for Linear or Non Linear Unknown Channels. *IEEE Trans. Communications* 42:2406–2413, July 1994.

[35] Robertson, P., and T. Worz. Bandwidth-Efficient Turbo Trellis-Coded Modulation Using Punctured Component Codes. *IEEE JSAC* 16(2):206–218, Feb. 1998.

[36] Gallager, R. G. *Low-Density Parity-Check Codes.* Cambridge, MA: MIT Press, 1963.

[37] Zyablov, V., and M. Pinkster. Estimation of the Error-correction Complexity of Gallager Low-density Codes. *Probl. Pered. Inform.*11:23–26, Jan. 1975.

[38] Margulis, G. A. Explicit Construction of Graphs without Short Cycles and Low Density Codes. *Combinatorica* 2(1):71–78, 1982.

[39] Tanner, R. A Recursive Approach to Low Complexity Codes. *IEEE Trans. Inform.Theory* IT-27:533–547, Sept. 1981.

[40] MacKay, D. J. C., and R. M. Neal. Near Shannon Limit Performance of Low Density Parity Check Codes. *Electron. Lett.* 32:1645–1646, Aug. 1996.

[41] Wiberg, N. Codes and Decoding on General Graphs. Dissertation on 440, Dept. Elect. Linköping Univ., Linköping, Sweden, 1996.

[42] Luby, M. G., M. Mitzenmacher, M. A. Shrokrollahi, D. Spielman, and V. Stemann. Practical Loss-Resilient Codes. *Proc. 29th Annu. ACM Symp. Theory of Computing*:150–159, 1997.

[43] Luby, M. G. M. Mitzenmacher, M. A. Shrokrollahi, D. Spielman, and V. Stemann. Practical Loss-Resilient Codes. *IEEE Trans. Inform. Theory* 47:569–584, Feb. 2001.

[44] Osman, O. Performance of Turbo Coding. Ph.D. thesis, Istanbul University, Turkey, 2003.

[45] Cirpan, H. A., and O. N. U•an. Blind Equalization of Quadrature Partial Response-Trellis Coded Modulated Signals in Rician Fading. *International Journal of Satellite Communications* 19:59–168, Mar.–April, 2001.

# Chapter 12

# Image Transmission Using Turbo Codes

Wireless multimedia is being discussed actively among researchers and developers. In a wireless environment, the transmission signal will suffer intersymbol interference (ISI) caused by multipath propagation. At high SNR, this existing ISI causes an irreducible "floor" for the BER [1]. The images that are to be transmitted over noisy channels are extremely sensitive to the bit errors that can severely degrade the quality of the image at the receiver. This necessitates the application of error control channel coding in the transmission. Powerful and effective channel coding is necessary for image transmission through a wireless environment. Recently turbo code has attracted enormous attention because of its remarkable error correcting capability [2,3]. Moreover, its iterative decoding scheme and soft channel information utilization are particularly suitable for compressed image transmission [4–8].

To compress and enhance the image in image transmission for better bit-error performance, adaptive Wiener-turbo system (AW-TS) and adaptive Wiener-turbo systems with JPEG and bit plane compressions (AW-TSwJBC) are introduced in [7].

## 12.1 Adaptive Wiener-Turbo System (AW-TS)

AW-TS is a combination of turbo codes and the two-dimensional (2-D) adaptive noise removal filtering. Each binary correspondence of the amplitude values of each pixel is grouped into bit planes. Then each bit of these bit planes is transmitted. At the receiver side, a combined structure, denoted as Wiener-turbo decoder is employed. The Wiener-turbo decoder is an iterative structure with a feedback link from the second turbo decoder and the Wiener filtering as in Figure 12.1. The decoding process continues iteratively until the desired output is obtained. The system consists of an image slicer, turbo encoder (transmitter), iterative 2-D adaptive filter, turbo decoder (receiver), and image combiner sections. The decoder employs two identical systematic recursive convolutional encoders (RSC) connected in parallel with an interleaver preceding the second recursive convolutional encoder. Both RSC encoders encode the information bits of the bit slices. The first encoder operates on the input bits in their original order, while the second one operates on the input bits as permuted by the interleaver. The decoding algorithm involves the joint estimation of two Markov processes, one for each constituent code. Because the two Markov processes are defined by the same set of data, the estimated data can be refined by sharing information between the two decoders in an iterative fashion. The output of one decoder can be used as *a priori* information by the other decoder. The iteration process is done until the outputs of the individual decoders are in the form of hard bit decisions. In this case, there is no advantage in sharing additional information.

The advantage of the neighborhood relation of pixels is taken into account in the AW-TS scheme, resulting in the improvement of image enhancement. Although each bit slice is transmitted in a serial way, the bit string is reassembled at the receiver in such a way that the original neighborhood matrix properties before the decoding process. Thus instead

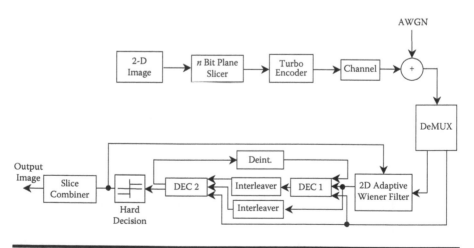

**Figure 12.1   AW-TS system model. (Copyright permission by IU-JEEE [7].)**

of a classic binary serial communication and decoding, in this scheme, the coordinates of the pixels are kept as in their original input data matrix.

Instead of transmission of the binary equivalence with the length of N of amplitude value of the pixels, each pixel value is mapped to a corresponding binary N-level; their binary correspondences are mapped regarding the quantization transmitted in bit slices as in Figure 12.2. The original image is partitioned into $2^N$ quantization levels, where N is denoted as the bit planes. Then each of the N-bit plane is coded by a turbo encoder and transmitted over an AWGN channel.

On the receiver side, bit planes are reassembled taking into account the neighborhood relationship of pixels in 2-D images. Each of the noisy bit-plane values of the image is evaluated iteratively by a combined equalization block that is composed of 2-D-adaptive noise-removal filtering and a turbo decoder.

In AW-TS, the data rate can be increased up to (N-1) times by only transmitting the most significant bit. Especially in quick search, such interesting results are given importance. Then the other bits can be transmitted to obtain more accurate 2-D images. Maximum resolution is obtained if all the bit slices from most significant to least significant are decoded at the receiver side without sacrificing bandwidth efficiency. Thus, our bit slicing also can be an efficient compression technique. As seen in Figures 12.3 and 12.4, the AW-TS output for each bit plane carries some part of the information.

At the receiver, after each bit slice is decoded and hard decision outputs are formed, all bit slice plane outputs are reassembled as the first bit from the first bit slice, second bit from the second bit slice, and the most significant bit from the last bit slice. Then these binary sequences are mapped to the corresponding amplitude value of the pixel. In the case of compression, due to the resolution, some but not all of the bits can be taken into account, and corresponding quantized amplitude values of the pixels found. Thus in AW-TS, both BER and image enhancement can

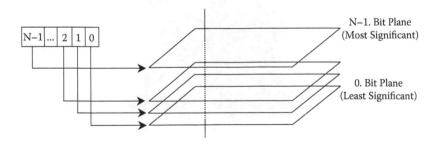

**Figure 12.2   N-bit plane decomposition of an image. (Copyright permission by IU-JEEE [7].)**

**Figure 12.3 Bit plane representation. (a) Original image; (b) 1.bit plane; (c) 2.bit plane; (d) 3.bit plane; (e) 4.bit plane. (Copyright permission by IU-JEEE [7].)**

**Figure 12.4** **The effects of the various slice combinations. (a) 0.bit plane+1.bit plane; (b) 0.bit plane+1.bit plane+2.bit plane; (c) 0.bit plane+1.bit plane+2.bit plane+3.bit plane; (d) 1.bit plane+2.bit plane; (e) 1.bit plane+2.bit plane+3.bit plane; (f) 2.bit plane+3.bit plane. (Copyright permission by IU-JEEE [7].)**

be achieved, compared to classic separation of turbo decoding and filtering. By sacrificing resolution, less memory storage is used and the data rate increased to (M-1) times by simply choosing any number of bit slices. In compression, resolution loss is a general case.

## 12.1.1 Bit Plane Slicing/Combining

For experimental results, a sample image with $150 \times 150$ pixel resolution with a 16-gray level is taken. The problem is to transmit the image that is extracted from the image resources, to the processing stations (ground stations, airplanes, ships, and so forth). While transmitting, the image is generally corrupted due to transmission conditions [2–5]. So it is necessary to minimize the noise effects.

Because binary turbo codes accept only binary inputs, the image must be converted to binary form first and then coded before transmission. The solution is using bit planes [7,8]. While the plane symbols are being transmitted, they are corrupted with a distorted channel, generally an AWGN.

Highlighting the contribution made to the total image appearance by a specific bit plays an important role for a AW-TS system. Data compression in the image processing area is another application of this technique. Imagine that the image is composed of N-bit planes, ranging from plane 0 for least significant bit to plane N-1 for the most significant bit. In terms of N-bit planes, plane 0 contains all the lowest order bits in the bytes comprising the pixels in the image and plane N-1 contains all the high-order bits. In other words, plane 0 is the least significant bit (LSB) and plane N-1 is the most significant bit (MSB). Figure 12.2 illustrates this.

This decomposition reveals that only some highest-order bits contain visually significant data. Also note that plane N-1 corresponds exactly with an image threshold at gray-level $2^{N-1}$. Figure 12.3 shows the various bit planes for the image for which all the pixels are represented by 4 bits (i.e., N = 3). In this case, we have $2^{3-1}$ bit planes. Note that only the three highest-order bits contain visually significant data. In [7,8], the image is sliced to 4 planes, i.e., each pixel in the image is represented by 4 bits (or 16 gray levels). Imagine that the image is composed of four 1-bit planes, ranging from plane 0 for the least significant bit to plane 3 for the most significant bit. In terms of 4-bit bytes, plane 0 contains all the lowest order bits among the bytes comprising the pixels in the image, while plane 3 contains all the high-order bits. Note that the most significant bit plane (3.bit plane) contains visually significant data. The other bit planes contribute the more subtle details of the image. Also plane 3 corresponds exactly with an image threshold at gray 8.

Bit plane combining is the reverse process of the slicing. The planes are recombined in order to reconstruct the image. However it is not necessary to consider all the slice contributions. Especially when the data rate is important, some planes can be ignored until the changes in gray level have an acceptable impact on the image. This approach will increase the data rate. Figure 12.4 shows how the combinations of the slices contribute to recovery of the image.

Transforming the image in a binary fashion before transmission is desirable. If the image had been considered without being sliced, then the neighborhood relationship would have been lost and it would have been pointless to have a Wiener filter at the receiver side; hence the performance of the proposed system would be the same as that of the other conventional techniques. When the image is sliced first then coded and transmitted, the neighborhood properties are evaluated. As a result, the noise effect is reduced and the performance significantly improved.

## 12.1.2 The Recursive Systematic Convolutional (RSC) Encoders

This section presents general information about RSC codes. Consider a half-rate RSC encoder with $M$ memory size. If the $d_k$ is an input at time $k$ the output $X_k$ is equal to

$$X_k = d_k \tag{12.1}$$

Remainder $r(D)$ can be found using feedback polynomial $g^{(0)}(D)$ and feedforward polynomial $g^{(1)}(D)$. The feedback variable is

$$r_i = d_k + \sum_{j=1}^{K} r_{k-j} g_j^{(0)} \tag{12.2}$$

and RSC encoder output $Y_k$, called parity data [5,6] is

$$Y_k = \sum_{j=0}^{K} r_{k-j} g_j^{(1)} \tag{12.3}$$

An RSC encoder with memory $M = 2$ and rate $R = 1/2$ with feedback polynomial $g^{(0)} = 7$ and feed forward polynomial $g^{(0)} = 5$ is illustrated in Figure 12.5, and it has a generator matrix

$$G(D) = \begin{bmatrix} 1 & \dfrac{1 + D + D^2}{1 + D^2} \end{bmatrix} \tag{12.4}$$

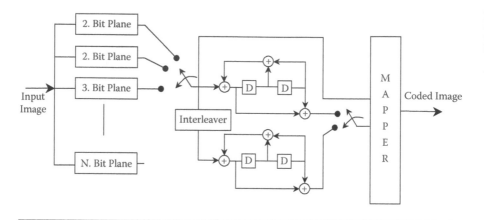

**Figure 12.5  AW-TS encoder. (Copyright permission by IU-JEEE [7].)**

## 12.1.3  *Wiener Filtering Applied to Bit Planes*

The goal of the Wiener filtering in Figure 12.1 is to obtain an estimate of the original bit plane from the degraded version. The degraded plane $g(n_1,n_2)$ can be represented by

$$g(n_1,n_2) = f(n_1,n_2) + v(n_1,n_2) \tag{12.5}$$

where $f(n_1,n_2)$ is the nice, under graded plane and $v(n_1,n_2)$ is the noise. Given the degraded plane $g(n_1,n_2)$ and some knowledge of the nature of $f(n_1,n_2)$ and $v(n_1,n_2)$, and some knowledge of the nature of $f(n_1,n_2)$ and $v(n_1,n_2)$, we want to come up with a function $h(n_1,n_2)$ that will output a good estimate of $f(n_1,n_2)$. This estimate is $p(n_1,n_2)$, and is defined by the following:

$$p(n_1,n_2) = g(n_1,n_2) * h(n_1,n_2)$$
$$P(w_1,w_2) = G(w_1,w_2)H(w_1,w_2) \tag{12.6}$$

The Wiener filter generates $h(n_1,n_2)$ that minimizes the mean square error that is defined by

$$E\{e^2(n_1,n_2)\} = E\{(g(n_1,n_2) - f(n_1,n_2))^2\} \tag{12.7}$$

This is a linear minimum mean square error (LMMSE) estimation problem, because it is a linear estimator, and the goal is to minimize the mean squared error between $g(n_1,n_2)$ and $f(n_1,n_2)$. This signal estimation problem can be solved using the orthogonality principle. It states that error

$e(n_1,n_2) = g(n_1,n_2) - f(n_1,n_2)$ is minimized by requiring that $e(n_1,n_2)$ be uncorrelated with any random variable. From the orthogonality principle,

$$E\{e(n_1,n_2)g^*(m_1,m_2)\} = 0 \quad \text{for all } (n_1,n_2) \text{ and } (m_1,m_2) \quad (12.8)$$

Then,

$$E\{f(n_1,n_2)g^*(m_1,m_2)\} = E\{(e(n_1,n_2) + p(n_1,n_2))g^*(m_1,m_2)\}$$

$$= E\{p(n_1,n_2)g^*(m_1,m_2)\}$$

$$= E\{(g(n_1,n_2)*b(n_1,n_2))g^*(m_1,m_2)\} \quad (12.9)$$

$$= \sum_{k_1=-\infty}^{\infty}\sum_{k_2=-\infty}^{\infty} b(k_1,k_2)E\{g(n_1-k_1,n_2-k_2)g^*(m_1,m_2)\}$$

Rewrite (12.9),

$$R_{fg}(n_1-m_1,n_2-m_2)$$

$$= \sum_{k_1=-\infty}^{\infty}\sum_{k_2=-\infty}^{\infty} b(k_1,k_2)R_g(n_1-k_1-m_1,n_2-m_2-m_2) \quad (12.10)$$

So

$$R_{fg}(n_1,n_2) = b(n_1,n_2)*R_g(n_1,n_2) \quad (12.11)$$

$$H(w_1,w_2) = \frac{P_{fg}(w_1,w_2)}{P_g(w_1,w_2)} \quad (12.12)$$

The filter in Equation (12.12) is called the non-causal Wiener filter. Suppose $f(n_1,n_2)$ is uncorrelated with $v(n_1,n_2)$

$$E\{f(n_1,n_2)v^*(m_1,m_2)\} = E\{f(n_1,n_2)\}E\{v^*(m_1,m_2)\} \quad (12.13)$$

Noting that $f(n_1,n_2)$ and $v(n_1,n_2)$ are a zero-mean process, we obtain:

$$R_{fg}(n_1,n_2) = R_f(n_1,n_2)$$

$$R_g(n_1,n_2) = R_f(n_1,n_2) + R_v(n_1,n_2) \quad (12.14)$$

and

$$P_{fg}(w_1, w_2) = P_f(w_1, w_2) \tag{12.15}$$

So,

$$H(w_1, w_2) = \frac{P_f(w_1, w_2)}{P_f(w_1, w_2) + P_v(w_1, w_2)} \tag{12.16}$$

If the additional constraints $f(n_1, n_2)$ and $v(n_1, n_2)$ are samples of Gaussian random field, then Equation (12.7) becomes an MMSE estimation problem, and the Wiener filter in Equation (12.16) becomes the optimal MMSE estimator. The Wiener filter in Equation (12.16) is a 0-phase filter. Since the power spectra $P(w_1, w_2)$ and $P_v(w_1, w_2)$ are real and nonnegative, $H(w_1, w_2)$ is also real and nonnegative. Therefore, the Wiener filter affects the spectral magnitude, but not the phase. By observing Equation (12.16), if $P_v(w_1, w_2)$ approaches 0, $H(w_1, w_2)$ will approach 1, indicating that the filter tends to preserve the high SNR frequency components. Let $P_v(w_1, w_2)$ approach infinity, $H(w_1, w_2)$ will approach 0, indicating that the filter tends to attenuate the low SNR frequency components.

The Wiener filter is applied to all planes and the additive noise on the bit-plane $v(n_1, n_2)$ is assumed to be zero mean and white with variance of $\sigma_v^2$. Let $f^j(n_1, n_2)$ show each plane from 0 to 3 (i.e., $j = 0,1,2,3$). Their power spectrums $P_v^j(w_1, w_2)$ are then given by $P_v^j(w_1, w_2) = (\sigma_v^j)^2$. Consider a small local region in which the plane $f^j(n_1, n_2)$ is assumed to be homogeneous. Within the local region, the plane $f^j(n_1, n_2)$ is modeled by

$$f^j(n_1, n_2) = m_f^j + \sigma_f^j w^j(n_1, n_2) \tag{12.17}$$

where $m_f^j$ and $\sigma_f^j$ are the local mean and standard deviation of $f^j(n_1, n_2)$ and $w^j(n_1, n_2)$ is zero-mean white noise with unit variance affecting each plane. Within the local region, then, the Wiener filter $H^j(w_1, w_2)$, $b^j(n_1, n_2)$ is given by

$$H^j(w_1, w_2) = \frac{P_f^j(w_1, w_2)}{P_f^j(w_1, w_2) + P_v^j(w_1, w_2)} = \frac{(\sigma_f^j)^2}{(\sigma_f^j)^2 + (\sigma_v^j)^2}; \tag{12.18}$$

$$j = 0,1,2,3$$

$$h^j(n_1,n_2) = \frac{(\sigma_f^j)^2}{(\sigma_f^j)^2 + (\sigma_v^j)^2}\delta(n_1,n_2); \quad j = 0,1,2,3 \quad (12.19)$$

Then, the restored planes $p^j(n_1,n_2)$ within the local region can be expressed as

$$p^j(n_1,n_2) = m_f{}^j + (g^j(n_1,n_2) - m_f{}^j) * \frac{(\sigma_f^j)^2}{(\sigma_f^j)^2 + (\sigma_v^j)^2}\delta(n_1,n_2)$$

$$(12.20)$$

$$= m_f{}^j + \frac{(\sigma_f^j)^2}{(\sigma_f^j)^2 + (\sigma_v^j)^2}(g^j(n_1,n_2) - m_f{}^j); \quad j = 0,1,2,3$$

If $m_f{}^j$ and $\sigma_f{}^j$ are updated at each symbol [9],

$$p^j(n_1,n_2) = m_f{}^j(n_1,n_2) + \qquad (12.21)$$

$$\frac{(\sigma_f^j)^2(n_1,n_2)}{(\sigma_f^j)^2(n_1,n_2) + (\sigma_v^j)^2(n_1,n_2)}(g^j(n_1,n_2) - m_f{}^j(n_1,n_2))$$

Experiments are performed with all the known degraded bit planes, which are degraded by adding from 2-DB to –3 dB SNR, and zero-mean white Gaussian noise. The resulting reconstructed images (by recombining the four planes) are shown in Figures 12.7(e)–12.12(e). The window size used to estimate the local mean and local variance is 5 by 5. From degraded planes, we can estimate $(\sigma_g^j)^2(n_1,n_2)$ can be estimated, and since $(\sigma_g^j)^2(n_1,n_2) = (\sigma_f^j)^2(n_1,n_2) + (\sigma_v^j)^2(n_1,n_2)$, Equation (12.21) changes to

$$p^j(n_1,n_2) = m_g{}^j(n_1,n_2) + \qquad (12.22)$$

$$\frac{(\sigma_g^j)^2(n_1,n_2) - (\sigma_v^j)^2(n_1,n_2)}{(\sigma_g^j)^2(n_1,n_2)}(g^j(n_1,n_2) - m_g(n_1,n_2));$$

$$j = 1,2,3,4$$

The generalized Wiener Equation (12.22) is modified as follows,

$$p_{i+1}^{(j)}(n_1,n_2) = m_{g_i}^{(j)}(n_1,n_2) +$$

$$\frac{(\sigma_{g_i}^{(j)})^2(n_1,n_2) - (\sigma_v^{(j)})^2(n_1,n_2)}{(\sigma_{g_i}^{(j)})^2(n_1,n_2)}(p_i^{(j)}(n_1,n_2) -$$

$$m_{g_i}^{(j)}(n_1,n_2)); \quad j = 1,2,3,4$$

(12.23)

The objective in [9] is to obtain better results at much lower SNRs. So, a MAP-based turbo-decoding algorithm is considered after the filtering process. In Equation (12.22), $p^j(n_1,n_2)$ ($j$ = 0,1,2,3) are taken as the degraded planes in the new inputs in the turbo decoders. The resulting output planes can be expressed as

$$p_{i+1}^{(0)}(n_1,n_2) = m_{g_i}^{(0)}(n_1,n_2) +$$

$$\frac{(\sigma_{g_i}^{(0)})^2(n_1,n_2) - (\sigma_v^{(0)})^2(n_1,n_2)}{(\sigma_{g_i}^{(0)})^2(n_1,n_2)}(p_i^{(0)}(n_1,n_2) -$$

$$m_{g_i}^{(0)}(n_1,n_2))$$

(12.24)

$$p_{i+1}^{(1)}(n_1,n_2) = m_{g_i}^{(1)}(n_1,n_2) +$$

$$\frac{(\sigma_{g_i}^{(1)})^2(n_1,n_2) - (\sigma_v^{(1)})^2(n_1,n_2)}{(\sigma_{g_i}^{(1)})^2(n_1,n_2)}(p_i^{(1)}(n_1,n_2) -$$

$$m_{g_i}^{(1)}(n_1,n_2))$$

$$p_{i+1}^{(2)}(n_1,n_2) = m_{g_i}^{(2)}(n_1,n_2) +$$

$$\frac{(\sigma_{g_i}^{(2)})^2(n_1,n_2) - (\sigma_v^{(2)})^2(n_1,n_2)}{(\sigma_{g_i}^{(2)})^2(n_1,n_2)}(p_i^{(2)}(n_1,n_2) -$$

$$m_{g_i}^{(2)}(n_1,n_2))$$

$$p_{i+1}^{(3)}(n_1,n_2) = m_{g_i}^{(3)}(n_1,n_2) +$$

$$\frac{(\sigma_{g_i}^{(3)})^2(n_1,n_2) - (\sigma_v^{(3)})^2(n_1,n_2)}{(\sigma_{g_i}^{(3)})^2(n_1,n_2)}(p_i^{(3)}(n_1,n_2) -$$

$$m_{g_i}^{(3)}(n_1,n_2))$$

where $i$ is the iteration index for each plane, between the Wiener process and turbo decoder (called the WT iteration index). $i$ can be taken from 1 to the desired number. If 1 is taken, the MAP-processed plane (at the output of the decoder) enters the Wiener filter only once. In this case, $P_1^j(n_1, n_2)$, $P_2^j(n_1, n_2)$ are chosen for each plane and the last iterated planes, i.e., $P_j^j(n_1, n_2)$, are taken into consideration for reconstruction. If the WT index is taken to be 0, the output of the decoders is not refiltered by Wiener. The WT index is taken at 1 in our study. The simulation results have shown that it is not necessary to take a very high value for the WT index (Figures 12.7–12.13).

## 12.1.4  Turbo Decoding Process

At the receiver side, before MAP-based decoding, noise reduction is done to obtain the highest probability of the symbols. The basic idea in the turbo scheme is for two or more *a posteriori* probability decoders to exchange soft information. One of the decoders calculates the *a posteriori* probability distribution of the bit planes and passes this information to the next decoder. The new decoder uses this information and computes its own version of the probability distribution. This exchange of information is called iteration. After a certain number of iterations, a decision is made at the second decoder. In each iteration, the probability of decoding in favor of the correct decision of each bit of the planes will improve. At the last iteration, the hard decision is made using the soft decision of the last decoder.

We consider the MAP approach for decoding as in Figure 12.6. The goal of the MAP algorithm is to find the *a posteriori* probability of each state transition, message bit, or code symbol produced by the underlying Markov process, given the noisy observation **y** [8–10].

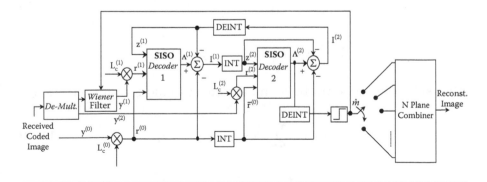

**Figure 12.6**  AW-TS system decoder. (Copyright permission by IU-JEEE [7].)

Once the *a posteriori* probabilities are calculated for all possible values of the desired quantity, a hard decision is made by taking the quantity with the highest probability. When used for turbo decoding, the MAP algorithm calculates the *a posteriori* probabilities of the message bits for the filtered planes, $P\left[m_i = 1 \mid y\right]$ and $P\left[m_i = 0 \mid y\right]$, which are then put into LLR form according to

$$\Lambda_k = \ln \frac{P\left[m_k = 1 \mid y\right]}{P\left[m_k = 0 \mid y\right]} \tag{12.25}$$

Before finding the *a posteriori* probabilities for the message bits, the MAP algorithm first finds the probability $P\left[s_i \to s_{i+1} \mid y\right]$ of each valid state transition given the noisy channel observation **y**. From the definition of conditional probability

$$P\left[s_i \to s_{i+1} \mid y\right] = \frac{P\left[s_i \to s_{i+1}, y\right]}{P\left[y\right]} \tag{12.26}$$

The properties of the Markov process can be used to partition the numerator as

$$P\left[s_i \to s_{i+1}, y\right] = \alpha(s_i)\gamma(s_i \to s_{i+1})\beta(s_{i+1}),$$

where

$$\alpha(s_i) = P\left[s_i, (y_0, \dots, y_{i-1})\right]$$

$$\gamma(s_i \to s_{i+1}) = P\left[s_{i+1}, y_i \mid s_i\right] \tag{12.27}$$

$$\beta(s_{i+1}) = P\left[(y_{i+1}, \dots, y_{L-1}) \mid s_{i+1}\right].$$

The probability $\alpha(s_i)$ can be found according to the forward recursion

$$\alpha(s_i) = \sum_{s_{i-1}} \alpha(s_{i-1})\gamma(s_{i-1} \to s_i) \tag{12.28}$$

The backward recursion $\beta(s_i)$ can be found as

$$\beta(s_i) = \sum_{s_{i-1}} \beta(s_{i-1})\gamma(s_i \rightarrow s_{i+1}) \tag{12.29}$$

Once the *a posteriori* probability of each state transition $P\left[s_i \rightarrow s_{i+1} \mid y\right]$ is found, the message bit probabilities can be found according to

$$P\left[m_i = 1 \mid y\right] = \sum_{S_1} P\left[s_i \rightarrow s_{i+1} \mid y\right], \tag{12.30}$$

And

$$P\left[m_i = 0 \mid y\right] = \sum_{S_0} P\left[s_i \rightarrow s_{i+1} \mid y\right]$$

where $S_1 = \{s_i \rightarrow s_{i+1} : m_i = 0\}$ is the set of all state transitions associated with a message bit of 0, and $S_0 = \{s_i \rightarrow s_{i+1} : m_i = 1\}$ is the set of all state transitions associated with a message bit of 1. The log-liklihood then becomes

$$\Lambda_i = \ln \frac{\displaystyle\sum_{S_0} \alpha(s_i)\gamma(s_i \rightarrow s_{i+1})\beta(s_{i+1})}{\displaystyle\sum_{S_1} \alpha(s_i)\gamma(s_i \rightarrow s_{i+1})\beta(s_{i+1})} \tag{12.31}$$

Now let

$$\tilde{\beta}(s_i) = \ln\beta(s_i) = \ln\left\{\sum_{s_{i+1}} \exp\left[\tilde{\beta}(s_{i+1}) + \tilde{\gamma}(s_i \rightarrow s_{i+1})\right]\right\} \tag{12.32}$$

Likewise, let

$$\tilde{\alpha}(s_i) = \ln\alpha(s_i) = \ln\left\{\sum_{s_{i-1}} \exp\left[\tilde{\alpha}(s_{i-1}) + \tilde{\gamma}(s_{i-1} \rightarrow s_i)\right]\right\} \tag{12.33}$$

So, LLR can be found as [11–13]:

$$
\begin{aligned}
\Lambda_i = \ln &\left\{ \sum_{S_1} \exp\left[ \tilde{\alpha}(s_i) + \tilde{\gamma}(s_i \to s_{i+1}) + \tilde{\beta}(s_{i+1}) \right] \right\} - \\
&\ln\left\{ \sum_{S_0} \exp\left[ \tilde{\alpha}(s_i) + \tilde{\gamma}(s_i \to s_{i+1}) + \tilde{\beta}(s_{i+1}) \right] \right\}
\end{aligned}
\tag{12.34}
$$

This soft bit information is de-interleaved and evaluated at the hard decision section.

## 12.1.5 Simulation Results of AW-TS System

In this subsection, simulation results illustrate the performance of the proposed (AW-TS) algorithm for transmitting the image. In [7,8], a 1/2 rate RSC encoder with an AWGN channel model is investigated. Here the generator matrix is $g$ = [111:101], a random interleaver is used and the frame size is chosen as N = 150, an iteration (between the Wiener filter and decoder) is taken 1. At first, the pixels of the image are converted to 16 gray levels and then sliced to four bit-planes. All planes are coded via a RSC encoder and a random interleaver. The coded planes are corrupted with SNR = 2-DB, 1 dB, 0 dB, –1 dB, –2-DB and –3 dB and transmitted. Figures 12.7–12.12 show the corrupted, traditional turbo-processed, well-known Wiener image processed, and reconstructed images via the AW-TS system

In AW-TS, there is an iterative feedback link between Wiener filtering and turbo decoder and process. The scheme employs a pixel-wise adaptive Wiener-based turbo decoder and uses statistics (mean and standard deviation of local image) of estimated values of the local neighborhood of each pixel. It gives extraordinarily satisfactory results for both BER and image enhancement performance for less than 2 dB SNR values, compared to separate application of a traditional turbo coding scheme and 2-D filtering. Compression can also be achieved by using AW-TS systems. In compression, less memory storage is used and it can increase the data rate up to N-1 times by simply choosing any number of bit slices, sacrificing resolution.

**Figure 12.7** Noisy 2-D image and various simulations for SNR = 2dB. (a) Corrupted image; (b) turbo processed; (c) Wiener processed image; (d) most significant part of AW-TS output; (e) AW-TS processed. (Copyright permission by IU-JEEE [7].)

**Figure 12.8    Noisy 2-D image and various simulation results for SNR = 1dB. (a) Corrupted image with SNR = 1dB; (b) turbo processed; (c) Wiener processed; (d) most significant part of AW-TS output; (e) AW-TS processed. (Copyright permission by IU-JEEE [7].)**

**Figure 12.9** **Noisy 2-D image and various simulations for SNR = 0dB. (a) Corrupted image with SNR = 0dB; (b) turbo processed; (c) Wiener processed; (d) most significant part of AW-TS output; (e)AW-TS processed. (Copyright permission by IU-JEEE [7].)**

**Figure 12.10** Noisy 2-D image and various simulations for SNR = −1dB. (a) Corrupted image with SNR = −1dB; (b) turbo processed; (c) Wiener processed; (d) most significant part of AW-TS output; (e) AW-TS processed. (Copyright permission by IU-JEEE [7].)

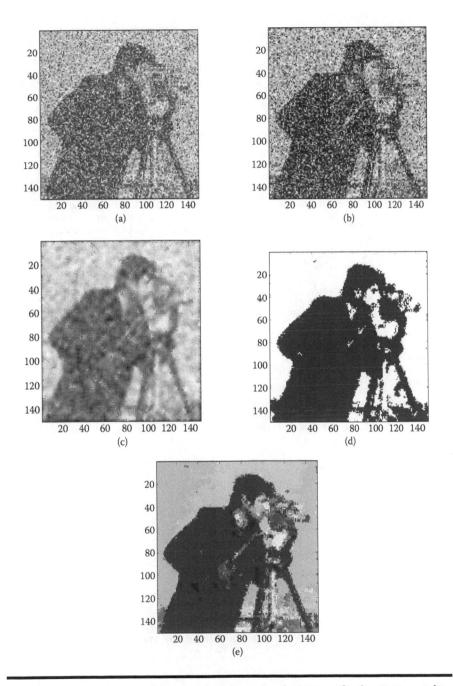

**Figure 12.11** Noisy 2-D image and various simulation results for SNR = –2dB. (a) Corrupted image with SNR = –2dB; (b) turbo processed; (c) Wiener processed; (d) most significant part of AW-TS output; (e) AW-TS processed. (Copyright permission by IU-JEEE [7].)

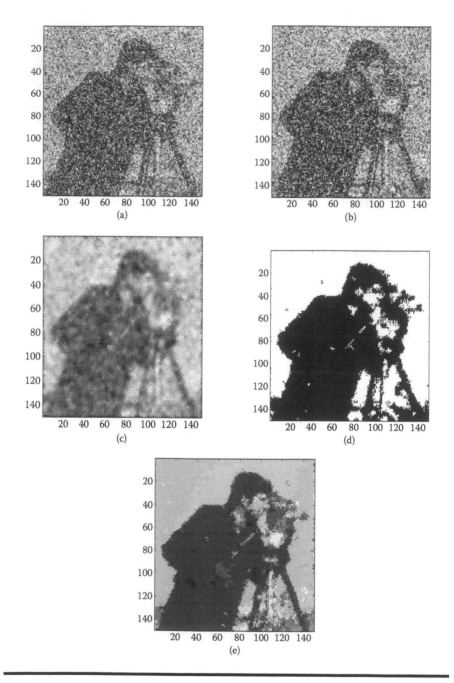

**Figure 12.12   Noisy 2-D image and various simulations for SNR = –3dB. (a) Corrupted image with SNR = –3dB; (b) turbo processed; (c) Wiener processed; (d) most significant part of AW-TS output; (e) AW-TS processed. (Copyright permission by IU-JEEE [7].)**

## 12.2 Adaptive Wiener-Turbo Systems with JPEG and Bit Plane Compressions (AW-TSwJBC)

To improve the unequal error protection and compression ratio of 2-D colored images over the wireless environment, we propose a new scheme denoted as adaptive Wiener-turbo systems with JPEG and bit plane compressions (AW-TSwJBC) [7]. In Figure 12.13, at the transmitter side of the AW-TSwJBC scheme, a 2-D colored image is passed through a color and bit planes slicer block. In this block, each pixel of the input image is partitioned up to three main color planes as red, green and blue (RGB), and each pixel of the color planes is sliced up to N binary bit planes that correspond to the binary representation of pixels. Thus depending on the importance of information knowledge of the input image, pixels of each color plane can be represented by a fewer number of bit planes. Then they are compressed by JPEG prior to turbo encoding. Hence, two consecutive compressions are achieved regarding the input image.

In 2-D images, information is mainly carried by neighbors of pixels [9–19]. Thus, in the AW-TSwJBC scheme, although each pixel of the image is transmitted in an aserial way, the bit string is reassembled at the receiver in such a way that the original neighborhood matrix relations are taken into consideration before the last decision process. Here, we benefit by

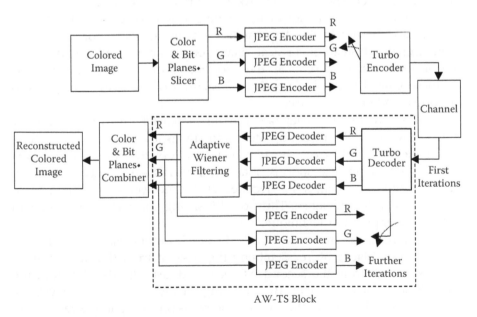

**Figure 12.13** **AW-TSwJBC system block diagram. (Copyright permission by IU-JEEE [7].)**

the neighborhood relation of pixels for each color plane by using a new iterative block, the adaptive Wiener–turbo scheme, which employs a turbo decoder, JPEG encoder/decoders, and adaptive Wiener filtering. After satisfactory results are observed, iterations are stopped, the 2-D image is passed through a color and bit planes combiner, and the reconstructed image is obtained.

Hence, in AW-TSwJBC scheme, turbo codes and various compression approaches, and image processing techniques are combined to obtain better bit-error performance and higher compression ratios with less data loss. In the considered approach, compression ratios can be changed according to the importance of the image to be transferred.

On the transmitter side, the AW-TSwJBC scheme uses bit plane slicing (BPS) [7], JPEG [8,9], and turbo codes [10], for compression and transmission improvement, respectively. The compression scenario reduces the significant amount of data required to reproduce the image at the receiver. In an AWGN environment, 2-D adaptive noise removal filtering is used at the receiver side to achieve better error performance. In the AW-TSwJBC scheme, each binary correspondence of the amplitude values of each pixel is grouped in bit planes. A combined structure, denoted as adaptive Wiener-turbo decoder is employed. The adaptive Wiener-turbo decoder is an iterative structure with a feedback link between the turbo decoder and Wiener filter. The decoding process works iteratively until the desired output is obtained. The advantage of neighborhood relation of pixels is taken into account in the AW-TSwJBC scheme to gain improved image enhancement. Instead of classical binary serial communication and decoding in the considered scheme, the coordinates of the pixels are kept as in their original input data matrix and each bit of the pixel is mapped to a corresponding slice matrix.

In AW-TSwJBC, the data rate can be increased up to (N) times, by transmitting only the most important and significant bit plane slice (BPS). Then the other BPSs can be transmitted to obtain more accurate 2-D images. Maximum resolution is obtained if all BPSs from most significant to least significant planes are decoded at the receiver side without sacrificing bandwidth efficiency. Thus, our bit slicing approach can also be an efficient compression technique.

At the receiver, after each BPS is decoded, output bits are mapped to the previous places by reassembling the first bit from the first bit slice, the second bit from the second bit slice, and lastly the last bit from the last bit slice for each pixel. By collecting all the pixels, the recovery of the image is achieved. In the case of compression owing to the resolution, some but not all of the BPSs can be taken into account. Thus, corresponding quantized amplitude values of the pixels are found. Resolution loss is a general case in compression, but by sacrificing resolution, less memory storage is used and the data rate is increased up to (N) times by simply choosing any number of bit slices. The AW-TSwJBC scheme [7] consists of a color and bit planes slicer, JPEG encoder, and turbo encoder in the transmitter side; the AWTS employs a turbo decoder, JPEG decoder, adaptive Wiener filtering, and color and bit planes combiners in the receiver side as shown in Figure 12.13. To increase the performance of the scheme at the receiver, JPEG encoders are applied as a feedback link between adaptive Wiener filtering and the turbo decoder.

On the transmitter side of the proposed scheme, the colored image is sliced into RGB planes and each of the RGB planes is separated to various bit planes. As an example, consider 8 bits for each color plane; thus, a total of 24 bit planes are encoded with JPEG encoders prior to running the turbo encoder. The turbo encoder employs two identical RSC encoders connected in parallel with an interleaver preceding the second recursive convolutional encoder. Both RSC encoders encode the information bits of the bit slices. The first encoder operates on the input bits in their original order, as the second one operates on the input bits as permuted by the interleaver. These bit planes are transmitted over the AWGN channel by the turbo encoder block.

On the receiver side, the turbo decoder, which consists of two component decoders concatenated in parallel, performs a suboptimal log-MAP algorithm. In order to perform the MAP algorithm, an accurate estimation of the channel conditions, i.e., the noise variance, is required. When the required numbers of iterations have been completed, the hard decision is made to obtain the estimated symbols at a second decoder. At the first iteration of the turbo decoding process, turbo-coded noisy symbols are used, but at the iterations following, the symbols are passed through an adaptive Wiener filtering block. There is a link between the turbo decoder and the adaptive Wiener filtering that contains JPEG encoders. After sufficient iterations, the estimated symbols are rearranged as RGB planes, decoded by a JPEG decoder and filtered with a Wiener filter. The filtered RGB planes are combined with a color and bit planes combiner and, finally, the reconstructed color image is observed.

## 12.2.1 AW-TSwJBC Decoding Process

In the first iteration process, only MAP-based decoding is considered. The goal of the MAP algorithm is to find the *a posteriori* probability of each state transition, message bit, or code symbol produced by the underlying Markov process, given the noisy observation $y$ [12,13]. Once the *a posteriori* probabilities are calculated for all possible values of the desired quantity, a hard decision is made by taking the quantity with the highest probability. After enough iteration is achieved, the soft bit information defined in

Equation (12.34) is de-interleaved and evaluated at the hard decision section for each RGB color plane. They are injected into the JPEG decoders and then passed on to the adaptive Wiener filtering via Equation (12.24). To enhance image quality in additional iterations, adaptive Wiener filtering outputs are fed into JPEG encoders. Then the JPEG encoders outputs are evaluated as new inputs of the turbo decoder block. When a satisfactory result is observed, a hard decision is found using Equation (12.34). Thus at the last iteration AW-TS outputs are combined in a color and bit planes combiner and the reconstructed colored image is obtained.

## 12.2.2 Simulation Results of AW-TSwJBC System

The scheme in [7] is experimentally evaluated for the transmission of the 500 × 500 test image over an AWGN channel. The first step of the transmission is based on partitioning the image into three RGB planes that are then sliced into 8 BPSs. So 3 × 8 = 24 planes are coded independently, each carrying some important and highly protected neighborhood relations to the receiver, enabling the adaptive Wiener filter to be used; hence, the entire performance of the system is improved.

Here, the encoder is a 1/2 rate RSC encoder. The generator matrix is g = [111:101]. The interleaver is in a random form. The iteration (between the Wiener filter and decoder) number is taken at 3. The performance of the proposed scheme is evaluated for various SNRs (i.e., SNR = 1dB, 2dB and 3dB). Figures 12.14–12.16 show the reconstructed images of the AW-TSwJBC system.

On the receiver side, 24 bit planes are reassembled by considering the neighborhood relationship of the pixels. Each of the noisy bit planes is evaluated iteratively by a combined block, which is composed of a 2-D adaptive noise-removal filter and a turbo decoder. In AW-TSwJBC, there is an iterative feedback link between the Wiener filter and turbo decoder. This scheme employs a plane-wise adaptive Wiener-based turbo decoder and uses statistics (mean and standard deviation of each plane) for the local neighborhood of each pixel to obtain the highest probability of the symbols.

Figures 12.14–12.16 show the re-assembled bit planes and the resulting image at the decoder output for SNR = 1, 2, and 3 dB values. In Figure 12.14(a), for each RGB plane, we have 8 BPS, two of them are transmitted, enabling 75-percent compression ratio, by BPS compression. The transmitted planes are then recompressed using a JPEG approach that is based on the discrete cosine transform (DCT) compression. The JPEG process involves first dividing the image into 8 × 8 pixel blocks. Each block's information is transformed to a frequency domain; all 1 × 1 low-frequency elements are transmitted and the others are discarded.

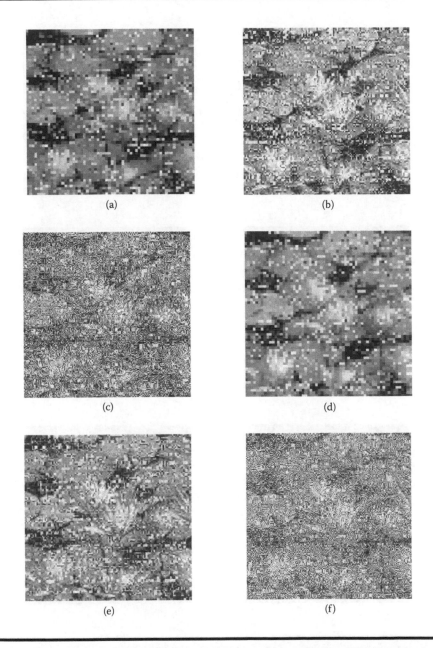

**Figure 12.14   Reconstructed Image for SNR = 1dB. (a) Total 96.875 compression (with 87.5 percent JPEG and 75 percent bit plane); (b) total 93.75 compression (with 75 percent JPEG and 75 percent bit plane); (c) total 87.5 compression (with 50 percent JPEG and 75 percent bit plane); (d) total 93.75 compression (with 87.5 percent JPEG and 50 percent bit plane); (e) total 87.5 compression (with 75 percent JPEG and 50 percent bit plane); (f) total 75 compression (with 50 percent JPEG and 50 percent bit plane). (Copyright permission by IU-JEEE [7].)**

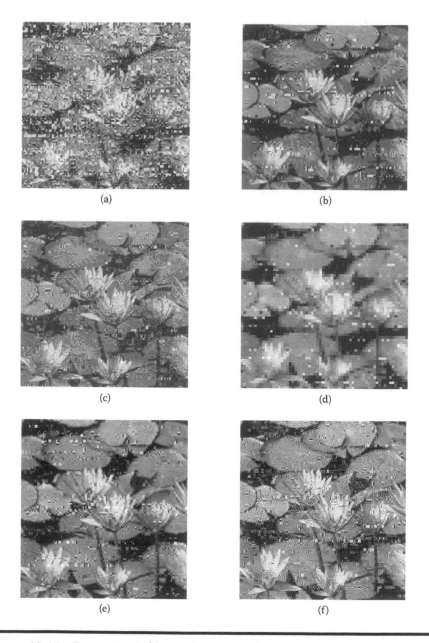

**Figure 12.15** **Reconstructed image for SNR = 2dB. (a) Total 96.875 compression (with 87.5 percent JPEG and 75 percent bit plane); (b) total 93.75 compression (with 75 percent JPEG and 75 percent bit plane); (c) total 87.5 compression (with 50 percent JPEG and 75 percent bit plane); (d) total 93.75 compression (with 87.5 percent JPEG and 50 percent bit plane); (e) total 87.5 compression (with 75 percent JPEG and 50 percent bit plane); (f) total 75 compression (with 50 percent JPEG and 50 percent bit plane). (Copyright permission by IU-JEEE [7].)**

Figure 12.16 Reconstructed image for SNR = 3dB. (a) Total 96.875 compression (with 87.5 percent JPEG and 75 percent bit plane); (b) total 93.75 compression (with 75 percent JPEG and 75 percent bit plane); (c) total 87.5 compression (with 50 percent JPEG and 75 percent bit plane); (d) total 93.75 compression (with 87.5 percent JPEG and 50 percent bit plane); (e) total 87.5 compression (with 75 percent JPEG and 50 percent bit plane); (f) total 75 compression (with 50 percent JPEG and 50 percent bit plane). (Copyright permission by IU-JEEE [7].)

Hence an 87.5-percent compression ratio is obtained by JPEG. In this case, some high-frequency elements that contain a lot of detail can be lost. The more these high-frequency elements are discarded, the smaller the resulting file and the lower the resolution of the reconstituted image. As a result, a 75-percent compression ratio for BPS and a 87.5 percent for JPEG give us 96.875 percent total compression.

In Figure 12.14 (b) these values are 75 percent for each compression (BPS and JPEG). In other words, two plane slices are used and $2 \times 2$ low-frequency elements of each $8 \times 8$ block; the others are thrown away. So we have 93.75 percent total compression. For Figure 12.14(c) these values are 75 percent for BPS and 50 percent for JPEG compression. Two plane slices are transmitted and $4 \times 4$ low-frequency elements of each $8 \times 8$ block; the others are thrown away. So 87.5 percent total compression is obtained. For Figure 12.14(d) these values are 50 percent for BPS and 87.5 percent for JPEG compression. Here, four plane slices are sent and $1 \times 1$ low-frequency elements of each $8 \times 8$ block; the others are thrown away. So 93.75 percent total compression is achieved. For Figure 12.14(e) these values are 50 percent for BPS and 75 percent for JPEG compression. In this case, four plane slices are transmitted and $2 \times 2$ low-frequency elements of each $8 \times 8$ block; the others are thrown away. So we have 87.5 percent total compression. Also, Figures 12.15 and 12.16 (a–f) are evaluated for 2- and 3-dB SNR values with the same compression ratios as in Figure 12.14.

A reliable and efficient compression–transmission system is explained for 2-D images using BPS and JPEG (for compression), and TC (for error correction) is called the AW-TSwJBC system. Traditional methods are time consuming, but the proposed method promises to speed up the process, enabling a better BER. In compression, less memory storage is used and the data rate is increased up to $N$ times by simply choosing any number of bit slices, sacrificing resolution. Hence, an AW-TSwJBC system is a compromise approach in 2-D image transmission, recovery of noisy signals, and image compression. Successive approximation is achieved in a high-SNR environment.

# References

[1] Chuang, J. C-I. The Effects of Time Delay Spread on Portable Radio Communication Channels with Digital Modulation. *IEEE Trans. SAC.*,5:879–889, 1987.
[2] Berrou, C., and A. Glavieux. Near Optimum Error Correcting Coding and Decoding: Turbo Codes. *IEEE Trans. Commun.* 44(10):1261–1271, 1996.

[3] U•an, O. N., K. Buyukatak, E. Gose, O. Osman, and N. Odabasioglu. Performance of Multilevel-Turbo Codes with Blind/non-Blind Equalization over WSSUS Multipath Channels. *International Journal of Communication Systems* 19(3):281–297, 2005.

[4] Thomos, N., N. V. Boulgouris, and M. G. Strintzis. Wireless Image Transmission Using Turbo Codes and Optimal Unequal Error Protection. *IEEE Transactions on Image Processing* 14(11):1890–1901, 2005.

[5] Yang, J., M. H. Lee, M. Jiang, and J. Y. Park. Robust Wireless Image Transmission based on Turbo-Coded OFDM. *IEEE Transactions on Consumer Electronics* 48(3):724–731, 2002.

[6] Han, Y. H., and J. J. Leou. Detection and Correction of Transmission Errors in JPEG Images. *IEEE Transactions on Circuits and Systems for Video Technology* 8(2):221–231, 1998.

[7] Buyukatak, K., E. Gose, O. N. U•an, and S. Kent. Adaptive Wiener-Turbo System (AW-TS) and Adaptive Wiener-Turbo Systems with JPEG & Bit Plane Compressions (AW-TSwJBC) Systems. *Istanbul University-Journal of Electrical & Electronics Engineering, IU-JEEE* 7 1:257–276, 2007.

[8] Buyukatak, K. Evaluation of 2-D Images for Turbo Codes. Ph.D. thesis, Istanbul Technical University, Institute of Science and Technology, Dec. 2004.

[9] Gonzales, R. C., and R. E. Woods. *Digital Image Processing*. Reading, MA: Addison-Wesley, 1992.

[10] Pennebaker, W.B., and J. L. Mitchell. *JPEG: Still Image Data Compression Standard*. New York: Van Nostrand Reinhold, 1993.

[11] CCITT Recommendation T.81. Digital Compression and Coding of Continuous-Tone Still Images, 1992.

[12] Valenti, M. C. *An Introduction to Turbo Codes*. Blacksburg, VA: Virginia Polytechnic Inst.& S.U.

[13] Hillery, A. D., and R. T. Chin. Iterative Wiener Filters for Image Restoration. *IEEE Trans. Signal Processing* 39(8):1892–1899,1991.

[14] Valenti, M. C. Iterative Detection and Decoding for Wireless Communications. *A Proposal for Current and Future Work toward Doctor of Philosophy degree,* 1998.

[15] Hageneauer, J. Iterative Decoding of Binary Block and Convolutional Codes. *IEEE Trans. Inform. Theory,* 42:429–445, 1996.

[16] Gross, W. J., and P. G. Gulak. Simplified MAP Algorithm Suitable for Implementation of Turbo decoders. *Electronic Letters* 34(16):1577–8, 1998.

[17] Buyukatak, K, E. Gose, O. N. U•an, S. Kent, and O. Osman. Channel Equalization and Noise Reduction based Turbo Codes. *Recent Advances on Space Technology,* 2003

[18] Gose, E., K. Buyukatak, O. Osman, O. N. U•an, and H. Pastaci. Performance of Turbo Decoding for Time-Varying Multipath Channels. *Recent Advances on Space Technology,* 2003

[19] Sklar, B. A Primer on Turbo Concepts. *IEEE Communications Magazine,* Dec. 1997.

# Index